Mechanics of
Nondestructive Testing

Mechanics of
Nondestructive Testing

Edited by
W. W. Stinchcomb
Virginia Polytechnic Institute and State University
Blacksburg, Virginia

Associate Editors:

J. C. Duke, Jr.
E. G. Henneke, II
K. L. Reifsnider
Virginia Polytechnic Institute and State University
Blacksburg, Virginia

Plenum Press · New York and London

Library of Congress Cataloging in Publication Data

Conference on the Mechanics of Nondestructive Testing, Virginia Tech, 1980.
    Mechanics of nondestructive testing.

    Papers from a conference held Sept. 10—12, 1980, sponsored by the Materials
Response Group, Engineering Science and Mechanics Dept., Virginia Polytechnic
Institute and State University.
    Includes index.
    1. Non-destructive testing—Congresses. I. Stinchcomb, W. W. II. Virginia Poly-
technic Institute and State University. Engineering Science and Mechanics Dept.
Materials Response Group. III. Title.
TA417.2.C66 1980   620.1'127   80-23808
ISBN 0-306-40567-9

Proceedings of the Conference on the Mechanics of Nondestructive Testing,
held at Virginia Polytechnic Institute and State University, Blacksburg, Virginia,
September 10—12, 1980.

© 1980 Plenum Press, New York
A Division of Plenum Publishing Corporation
227 West 17th Street, New York, N.Y. 10011

Printed in the United States of America

PREFACE

The synergism of the mechanics of nondestructive testing and the mechanics of materials response has great potential value in an era of rapid development of new materials and new applications for conventional materials. The two areas are closely related and an advance in one area often leads to an advance in the other. As our understanding of basic principles increases, nondestructive testing is outgrowing the image of "black box techniques" and is rapidly becoming a legitimate technical area of science and engineering. At the present time, however, an understanding of the mechanics of nondestructive testing is lagging behind other advances in the field. The key to further development in the mechanics of nondestructive testing lies in the mechanics of the phenomena or response being investigated — a better understanding of materials response suggests better nondestructive test methods to investigate the response which, in turn, advances our understanding of materials response, and so on.

With this approach in mind, the Materials Response Group of the Engineering Science and Mechanics Department at Virginia Polytechnic Institute and State University hosted a Conference on the Mechanics of Nondestructive Testing on September 10 through 12, 1980. Sponsors of the conference were the Army Research Office, the National Science Foundation, and the Engineering Science and Mechanics Department. The objective of the conference was to promote an understanding of the mechanics of nondestructive testing as related to the evaluation of materials response. The papers in this volume cover many of the frequently used NDT methods and introduce several new and innovative ideas in the field. Five overview papers present the general philosophy of optical techniques, analysis of acoustic emission, electromagnetic methods, stiffness measurements, and ultrasonic techniques. Collectively, the papers seek to develop relationships between nondestructive methods of investigation and the response of materials.

Our sincere appreciation is extended to all who worked so hard and contributed to the success of the conference: the authors and participants, Dr. Daniel Frederick, Head of the Engineering Science

v

and Mechanics Department, Dean Paul Torgersen, Dean of the College of
Engineering, Dr. Clifford Astill, the National Science Foundation,
Dr. Fred Schmiedeshoff, the Army Research Office, Dr. Ralph Siu, and
Dr. Richard Harshberger and the staff of the Donaldson Brown Con-
tinuing Education Center at Virginia Tech. A very special recog-
nition and note of gratitude is expressed to Mrs. Phyllis Schmidt and
Mrs. Betty Eaton who worked with patience and dedication in preparing
the manuscripts for publication.

                                        Wayne W. Stinchcomb, Editor
                                        and
                                        John C. Duke,
                                        Edmund G. Henneke, II,
                                        Kenneth L. Reifsnider,
                                                    Associate Editors

September, 1980
Virginia Polytechnic Institute
     and State University
Blacksburg, Virginia

CONTENTS

SESSION III
Defect and Flaw Characterization
Robert Crane, Chairman

SESSION IV
Material Damage-Initiation and Growth: Life Prediction
Francis Chang, Chairman

CONTENTS

OPTICAL INTERFERENCE FOR DEFORMATION MEASUREMENTS--
CLASSICAL, HOLOGRAPHIC AND MOIRÉ INTERFEROMETRY

Daniel Post

Virginia Polytechnic Institute and State University
Blacksburg, Virginia 24061

INTRODUCTION

Interferometry is a powerful tool. It enables measurements
of surface topography, dimensional changes and deformations of
solid bodies, all in the sub-wavelength sensitivity range. This
article is a tutorial treatment of the subject. Its purpose is
to introduce modern interferometric techniques of engineering
measurements; and to explain them in sufficient detail to generate a comfortable association with the subject. Ultimately,
the mission is to foster understanding and appreciation in order
to inspire increased utilization and further development.

The fundamentals of interferometry are easy to understand.
The rather diverse implementations, including holographic and
moiré interferometry, fit under a common umbrella. Only a few
basic concepts and a few equations are involved and these will be
reviewed in sufficient depth for practical usefulness. Interferometry is easy to apply to many important measurements problems.
The name has long been associated with delicate instrumentation
and masterful laboratory technique. These images no longer
represent many applications, and they are grossly relaxed in most
others. More than anything else, the substitution of lasers in
place of classical light sources has simplified the technology.

The fundamental ideas and relationships will be reviewed
first. Classical interferometry will be described, but with the
admission of lasers to the classical technology. Then, holography
and holographic interferometry will be explained. These techniques are especially useful for out-of-plane measurements, e.g.,
surface topography, change of specimen length, and displacements

1

normal to the specimen surface.

Moiré interferometry will be presented last. This is an emerging
technology to measure in-plane displacements, i.e., displacements
in the plane of the specimen surface, with high sensitivity and
accuracy. While placed last by historical chronology, it holds
the promise of a powerful NDT technique for observation and
measurement of material integrity and structural performance.

        Familiar codes will be used in the diagrams. It is es-
pecially important to remember that a sketch of an eye (as in
Fig. 9) represents any observing system, such as a human eye, a
camera or a video receiver. In addition, a ray of light in a
diagram (as in Fig. 2) represents the beam of light that contains
the ray; when a reader sees a simple ray, he/she should always
picture the full beam and the wave trains and wavefronts traveling
in the beam. Always.

FUNDAMENTALS

Our Model of Light

        The wave theory of light is sufficient to explain all the
characteristics we use. A parallel beam of light emitted in the
z-direction is depicted (at a given instant) as a train of regularly spaced disturbances that vary with z as

$$A = a \cos 2\pi \frac{z}{\lambda} \tag{1}$$

Symbol A describes the amplitude or strength of the disturbance,
which is usually viewed as the strength of an electro-magnetic
field at a point in space. Coefficient a is a constant. The
field strength varies harmonically along z, where the distance
between neighboring maxima is λ, called the wavelength. Length
z is not endless--for that would represent an eternal light--
but it is very long compared to λ; in the case of laser light,
length z of the wave train may be a million wavelengths or more.
Many such wave trains exist simultaneously in a beam of light.

        However, the wave train is not stationary. It travels or
propagates through space with a very high constant velocity C.
At any fixed point along the path of the wave train, the disturbance is a periodic variation of field strength. Field strength
varies through one full cycle in the brief time interval λ/C
(seconds/cycle); its frequency ω is C/λ (cycles/second). During
the passage of the wave train through any fixed point $z = z_o$, the
light disturbance varies with time t as

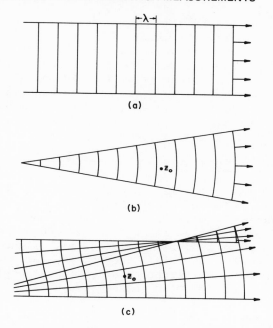

Fig. 1   Wavefronts for (a) a parallel beam
of light; (b) a spherical or conical
beam; and (c) an irregular beam. λ
is wavelength, or distance between
adjacent wavefronts.

$$A = a \cos 2\pi \frac{C}{\lambda} t = a \cos 2\pi\omega t \qquad (2)$$

The variable $2\pi\omega t$ is called the *phase* of the disturbance.
Along any cross-section of a parallel beam of light, the phase
is constant; in other words, the field strength is the same at
all points in the cross-section.

We will define a *wavefront* as any surface along which the
field strength is maximum, i.e., along which

$$\omega t = 0,1,2,3,--,k,--- \qquad (3)$$

where k is an integer. In a parallel beam, the wavefronts are
always plane cross-sections of the beam, as illustrated in Fig. 1a.
These move along the length of the beam with the speed of light.

In a conical beam (Fig. 1b), the wavefronts are spherical
and their radius of curvature increases with distance from the
origin. The shape of a wavefront changes constantly as it propagates through space, but all wavefronts in a wave train have a
fixed shape as they cross an arbitrary fixed point z_o in their path.

Beams may acquire irregular shapes (Fig. 1c), for example by reflection by an irregular mirror. Again the wavefronts constantly change shape as they propagate, and in extreme cases they may even fold and become double-valued as shown. This is a consequence of the rule that light propagates in the direction of the *wave normal*, that is, in the direction perpendicular to the wavefront.* As before, every wavefront in the beam has the same unique shape as it propagates across a fixed point z_o. Wavefronts of irregular shape are called *warped* wavefronts.

More Facts and Definitions

Light trains propagate with an astounding velocity of about 3×10^8 m/sec or 186,000 miles/sec in free space. The visible spectrum emcompasses the wavelength range from about 400 nm (i.e. nanometers or 10^{-9}m) for blue light to about 700 nm for red light. The inexpensive popular helium-neon laser emits light of wavelength 632.8 nm (24.91×10^{-6} in.) and the popular argon laser can be adjusted for several different colors, with strongest emissions at 514.5 nm (green) and 488.0 nm (blue-green).

The visible spectrum is only a tiny portion of the electro-magnetic spectrum in which wavelengths extend at least from 10^{-14} m to 10^3 m, a span of more than 17 decades. All the phenomena discussed here--interference, diffraction, etc.--apply for the entire electro-magnetic spectrum.

When a wave train crosses a fixed point in space, its frequency of oscillation $\omega = C/\lambda$ is about 6×10^{14} cycles/sec for visible light. No meters can detect individual cycles in this frequency range. Instead, receivers like the eye, photographic film and photoelectric cells respond to time-integrated values of energy and power. Power (or the rate of energy flow) per unit cross-sectional area of a beam of light is called its *intensity*. Energy intercepted by a unit area of a receiver is intensity multiplied by exposure time.

It is shown in electromagnetic theory that the intensity of light of amplitude $A = a \cos 2\pi\omega t$ is

$$I = a^2 \tag{4}$$

Thus, intensity is a time-averaged quantity and it is *independent of the frequency and phase* of the light beam.

The velocity C' of light propagation in a transparent material is less than its velocity in free space. The relation-

---

*This is always true for light propagation in homogeneous media.

ship is

$$C' = \frac{C}{n}$$

where n has a value for each material; n is a material property
called *index of refraction* or *refractive index*. For gasses, n
is only slightly greater than unity; for liquids, values between
1.3 and 1.5 are common; for solids, values between 1.5 and 1.7
are common.

The time required for a wavefront to travel from a point *a*
in its path to a point *b* depends not only upon the distance between
the points, but also upon the indicies of refraction of materials
in the path. If transparent materials of different refractive
index, n_1, n_2, n_3,---, lie in the path and if the actual path
length through each material is ℓ_1, ℓ_2, ℓ_3,---, respectively, the
transit time is

$$t = \frac{1}{C} (n_1\ell_1 + n_2 \ell_2 + n_3\ell_3 + ---).$$

Optical path length (OPL) is defined as the distance a beam would
travel in free space (vacuum) in the same time as required to
travel through the actual materials. Thus,

$$OPL = Ct = n_1\ell_1 + n_2\ell_2 + n_2\ell_2 + n_3\ell_3 ---. \tag{5}$$

As a corollary, the number of wave cycles experienced by the light
in the real path is identical to the number of cycles in an
equivalent OPL in free space.

Sometimes we are interested in *polarization* of the light.
This relates to the electromagnetic disturbance described by
Eq. 1, which shows A as a function of z, but independent of x
and y. This is a simplification of the facts, although it models
unpolarized light very well. With polarized light, the field
strength is unidirectional, say in the y-direction, and the disturbance is written

$$A_y = a \cos 2\pi \frac{z}{\lambda} . \tag{1a}$$

Then, field strength in the x-direction is zero, or

$$A_x = 0 \tag{1b}$$

which means there is no light with polarization parallel to x.

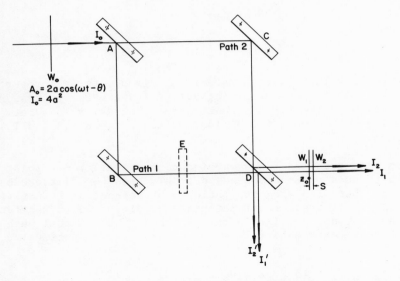

Fig. 2   Parallel beams I_1 and I_2 combine to produce con-
          structive or destructive interference, depending
          upon their path difference, or wavefront separa-
          tion, S.

INTERFERENCE OF LIGHT

The superposition of two *coherent* beams of light is called
interference. For our purposes, coherent beams are made up of
long wave trains; the intensity of each wave train is split into
two parts, and each component is coherent with respect to the
other--i.e., the ratio of their amplitudes (Eq. 2) is independent
of time. This infers that wavelength λ must have a unique value
in a coherent beam.

Pure Two-Beam Interference

In a parallel beam, let a wave train be split into two parts,
each with half the original intensity. Let the two parts travel
separate paths and then recombine into a common path. A means
of doing this is illustrated schematically in Fig. 2. Elements
A and D are *beam-splitters* or *half-mirrors*--mirrors that reflect
50% of the incident intensity and transmit the remainder;* B and
C are fully reflective mirrors. It is important to note that
beam-splitters are specified by their reflected and transmitted
intensities, not amplitudes.

---

*This assumes absorption is negligible, which is true for multi-
layer dieletric mirror coatings; ultra-thin metallic mirror coatings
are semi-transparent, but have substantial absorption. Both types
are produced commercially and routinely. Partial mirrors can be
made with any ratio of reflectance to transmittance.

An incident wave train with wavefront W_0 is divided at A and component trains of half the original intensity travel paths ABD and ACD. At D, each of these are divided again, such that components I_1 and I_2 emerge horizontally and I_1' and I_2' emerge vertically. We will be concerned with the specific wavefront W_0 as the wave train travels through the system. After division at mirrors A and D, components of W_0 emerge in the horizontal beam as W_1 and W_2.

Let the mirrors be adjusted for angles of exactly 45° and the path lengths adjusted for ABD = ACD. Then, optical path lengths of I_1 and I_2 are identical and wavefronts W_1 and W_2 are coincident.

Let a uniform transparent plate E be inserted in the system to increase the optical path length of path 1 by a length S. Consequently, a longer time is required for W_1 to negotiate its path and it emerges behind W_2. When W_1 reaches an arbitrary point z_0, its mate W_2 has traveled further by S or by S/λ cycles or by $2\pi S/\lambda$ radians of phase difference. If at z_0 the amplitude of wave train I_1 is assumed to vary as

$$A_1 = a \cos 2\pi\omega t, \tag{2a}$$

then the phase of wave train I_2 must be less by $2\pi S/\lambda$ when it crosses z_0, and the amplitude of I_2 must be

$$A_2 = a \cos 2\pi(\omega t - S/\lambda) \tag{2b}$$

Since the phase difference is the same for all time during the passage of the wave train (and for every other wave train through the system), the resultant amplitude of the light is

$$A = A_1 + A_2 = K \cos 2\pi(\omega t - \phi)$$

where

$$K = [2a^2(1 + \cos 2\pi S/\lambda)]^{1/2}$$

By Eq. 4, the resultant intensity of the recombined wave trains is K^2 or

$$I = 2a^2 (1 + \cos 2\pi S/\lambda) = 4a^2 \cos^2 \pi S/\lambda \tag{6}$$

This is a fundamental relationship, called the intensity distribution of pure two-beam interference. Intensity I applies to *every* case of division into two equal beams and their subsequent recombination. Equation 6 shows that the intensity of the combined beams varies cyclically from maximum to zero to maximum as a

function of the difference S of optical path lengths traveled
by the two beams. Why is it important? We can measure S if we
can count the cycles of intensity change.

Constructive interference occurs when the intensity is maximum, i.e. when S is an integral number of wavelengths, or S/λ =
0,1,2,3,---. *Destructive* interference corresponds to minimum
intensity, I_{min}, when $S = \lambda/2$ is an integral number of wavelengths,
or S/λ = 1/2, 3/2, 5/2,---. The present example is called *pure*
two-beam interference, since the interfering beams have equal
strengths and destructive interference is complete, producing
I_{min} = 0.

Games with Conservation

When the wavefront under consideration in Fig. 2 was positioned at W_o, the amplitude of the wave train was

$$A_0 = 2a \cos 2\pi(\omega t - \frac{OPL_1}{\lambda})$$

where OPL_1 is the optical path length of path 1 between the positions of W_o and W_1. The original intensity of the wave train was
thus

$$I_0 = 4a^2.$$

Assuming mirrors A and D each have 50% reflectance and 50% transmittance, and mirrors B and C have 100% reflectance, the intensities I_1, I_2, I_1', I_2' are each attenuated by 0.5^2 and are each
equal to $0.5^2 (4a^2) = a^2$. All of this is consistent with Eqs.
2a and 2b, which require $I_1 = I_2$.

However, Eq. 6 shows that $I = I_1 + I_2$ can be as high as $4a^2$,
which itself is equal to the original intensity I_0. Can we siphon
off energy from I_1' and I_2' and solve the potential U.S. energy
crisis?

Of course not! When $I = 4a^2$, interference of I_1' and I_2'
must give $I' = 0$. In fact, I and I' are always complimentary,
that is,

$$I' = 4a^2 \sin^2\pi S/\lambda$$

This is because a wave train experiences a phase change of $\pi/2$
upon reflection; since I_1' experiences three reflections while I_2'
experiences only one, an extra net retardation of π appears in
the amplitude summation. If not for this detail, we could create
energy!

Impure Two-Beam Interference

Suppose mirror A in Fig. 2 is a half-mirror, but mirror D has reflectance R and transmittance T. Then, intensities approaching D in paths 1 and 2 would each be $2a^2$ and I_1 and I_2 would be $2a^2T$ and $2a^2R$, respectively. Their amplitudes would be expressed very much like Eqs. 2a and 2b, namely

$$A_1 = (2a^2T)^{1/2} \cos 2\pi\omega t = a_1 \cos 2\pi\omega t \tag{2c}$$

and

$$A_2 = (2a^2R)^{1/2} \cos 2\pi(\omega t - S/\lambda) = a_2 \cos 2\pi(\omega t - S/\lambda) \tag{2d}$$

Like before, superposition or interference yields

$$A = A_1 + A_2 = K' \cos 2\pi(\omega t - \phi)$$

where

$$K' = (a_1{}^2 + a_2{}^2 + 2a_1a_2 \cos 2\pi S/\lambda)^{1/2}$$

and the resultant intensity is

$$I = (K')^2 = a_1{}^2 + a_2{}^2 + 2a_1a_2 \cos 2\pi S/\lambda \tag{7}$$

This is the general expression for interference, or superposition of any two coherent wave trains.

As an example, let T = 0.2 (20% transmittance) and R = 0.8. Equation 7 becomes

$$I = 2a^2(1 + 0.8 \cos 2\pi S/\lambda) \tag{7a}$$

Maximum intensity, I_{max}, occurs when the cosine term is one, and I_{min} occurs when it equals minus one. Thus

$$I_{max} = 3.6a^2; \quad I_{min} = 0.4a^2$$

and

$$I_{max} : I_{min} = 9:1$$

Interference of beams of unequal strengths is impure because destructive interference is incomplete, or $I_{min} \neq 0$. Still, the cyclic changes of intensity may be highly visible and easily counted.

TABLE I
TWO—BEAM INTERFERENCE FOR COHERENT BEAMS OF UNEQUAL INTENSITIES

Case	2-Beam Input		Output-Result of Coherent Summation	
	$I_1 = a_1^2$	$I_2 = a_2^2$	$\dfrac{I_{max}}{I_{min}}$	Contrast,%
a	1	1	∞	100
b	1	2	34	97
c	1	4	9	89
d	1	9	4	75
e	1	16	2.8	64
f	1	25	2.2	56
g	1	100	1.5	33

Table I shows this quality for different pairs of input strengths; the output is obtained from Eq. 7 and tabulated in terms of I_{max}/I_{min} and *contrast*, where

$$\%\text{contrast} = \frac{I_{max} - I_{min}}{I_{max}} \times 100\% \tag{8}$$

Resultant intensities are plotted in Fig. 3 as a function of path difference $N = S/\lambda$. In the example above, intensities of the two interfering beams were in the ratio of 4:1 and still contrast was very good, viz. 89%. It is amazing but true that superposition of two beams with 100:1 intensity ratio produces readily discernable interference with contrast of 33%.

FRINGE PATTERNS

The apparatus sketched in Fig. 4 is the same as that of Fig. 2, except uniform plate E is replaced by a transparent wedge F. A wedge merely bends a parallel beam of light, so that it emerges as a parallel beam that deviates from the original direction by an angle that we will call 2α. The angular deviation results from the linear variation of OPL from apex to the base of the wedge. As always (in free space and in isotropic media), its wavefront W_1 is perpendicular to the path of the beam.

Again let W_1 and W_2 be wavefronts that originated from division of wavefront W_0. Consequently, W_1 and W_2 have identical phase. Where S is an integral number of wavelengths, interference is constructive; midway between these locations interference is destructive.

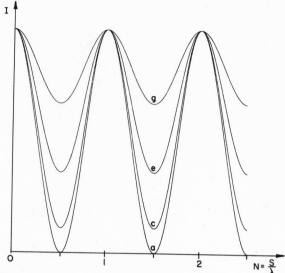

Fig. 3   Intensity varies cyclically with path difference
S.   Curves correpsond to Table I.

The analysis is aided by Fig. 5, which depicts wavefronts
W_1 and W_2, and neighboring wavefronts in the two wave trains.
Separation between wavefronts in each train is λ. Relative to
Figure 4, angle 2α is exaggerated and only small values of S
are indicated; however, the following argument applies for any
α between 0 and 90° and for any S--except S must be small compared to the length of the wave train.

The harmonic curves with ordinates labeled A_1 and A_2 represent the amplitude of the two wave trains in space at a given
instant of time. Phase ϕ_1 and ϕ_2 are identical since they represent wavefronts of common origin. Consequently, when W_1 crosses
point a, W_2 already advanced 5 wavelengths beyond a. Similarly
at b, where the optical path lengths traveled by the two wave
trains again differ by 5 wavelengths. This is true for every point
along a-c, so Eq. 6 prevails at all these points, where $S/\lambda = 5$
in the equation. The result is constructive interference along a-c.

By the same argument, along d-e the difference S of optical
path lengths traveled by the two wave trains is 4.5 wavelengths,
or $S/\lambda = 4.5$; along q-r, $S/\lambda = 4.0$. This explains the formation
of constructive and destructive interference at the given instant,
but why is the effect persistent, or independent of time?

At a later instant the two wave trains have advanced, but
their velocities are identical and the distances travelled are
identical. Therefore, the phase difference at point a remains

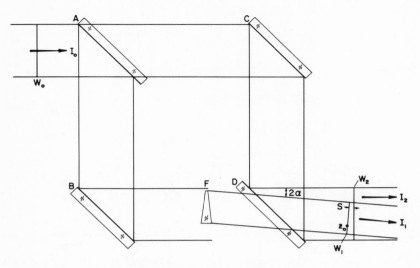

Fig. 4   Wavefront W_0 is divided and its components
         W_1 and W_2 travel paths that differ in length
         by S, where S is a variable.

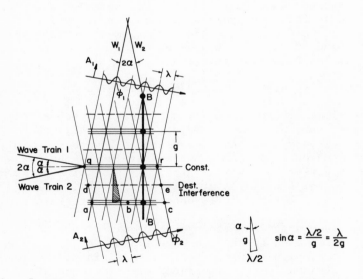

Fig. 5   Wave trains of Amplitudes A_1 and A_2 combine
         to produce steady-state constructive and de-
         structive interference.

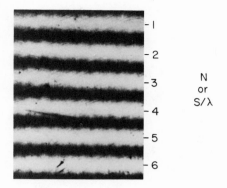

Fig. 6   Interference fringe pattern formed on photo-
         graphic plate B-B in Fig. 5.

5 cycles, and the phase difference at all other points remains
unchanged. A steady-state condition of constructive and destructive interference is formed.

Furthermore, the wavefronts of Fig. 5 are not merely lines
as shown, but they are planes lying perpendicular to the diagram.
Constructive and destructive interference occurs in a series of
parallel planes. One may think of these as parallel walls of
interference in space, always lying perpendicular to the bisector
of wavefronts W_1 and W_3 and always exhibiting an intensity distribution of Fig. 3.

A fundamental relationship defines the distance g between
adjacent walls of interference. From the shaded triangle, whose
hypotenuse is g and short leg is $\lambda/2$, we find

$$\sin \alpha = \frac{\lambda/2}{g} \tag{9}$$

Like Eq. 6 or 7, this relationship pertains to every case of
optical interference, and it will arise repeatedly in our analyses.

If we interpose a photographic plate along B-B, it would be
exposed in the zones of constructive interference and remain unexposed in the zones of destructive interference. The result
would be similar to Fig. 6, which is a positive print and shows
destructive interference as black, or the absence of light.
Bright bands, or *fringes* as they are called, represent zones of

$$S/\lambda = N = 0,1,2,3,--- \tag{10}$$

where N is called the *interference fringe order,* or simply
fringe order. The array of fringes in the figure is called a
fringe pattern.

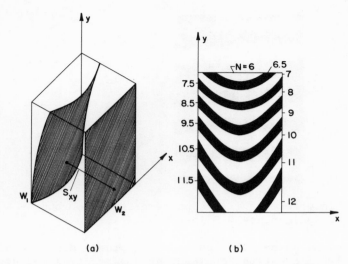

Fig. 7 (a) For warped wavefronts, separation S
           varies with x and y. (b) Corresponding
           fringe pattern is a contour map of S_{xy}
           with contour interval equal to wavelength λ.

Equation 10 shows that the fringe pattern is a contour map
of the separation S between wavefronts W_1 and W_2, where the contour interval is one wavelength, λ. (S is measured in the direction of travel of either wave train.)

Warped Wavefronts

Let us refer again to Fig. 4 and replace the wedge with
another transparent body whose surfaces are not flat, but de-
viate from flatness in a smooth continuous manner. As a result,
wavefront W_1 becomes smoothly warped, e.g., in the manner
sketched in Fig. 7a. Reference wavefront W_2 remains plane.
Now, interference between the two wave trains produces the
nonlinear pattern of Fig. 7b, which is a contour map of separation S_{xy} between the two wavefronts.

It should be evident that Eqs. 9 and 10 are both operative.
The warped wavefront is generated because angle α varies continuously with x and y. As a result, wavefront separation S and
fringe space g both vary systematically with x and y.

Along any continous bright contour, S_{xy}/λ has the constant
value given by fringe order N, where N is an integer. Similarly
for dark contours and also for continuous contours of any

intermediate intensity, so N assumes all values, integer, half-order, and intermediate values.

Since W_1 is smoothly warped, fringe orders must vary in a smooth manner. Fringes must be smooth contours and the changes of fringe order between adjacent fringes must be either 1 or zero.

Interferometry utilized in non-destructive testing will often involve only one warped surface. However, the interpretation is the same even if both beams have warped wavefronts. In any case, the resultant fringe pattern is a contour map of the separation S_{xy} between wavefronts W_1 and W_2, that is, between wavefronts of equal phase in the two beams.

THE CAMERA

One more idea is important before we discuss practical techniques. Referring again to Fig. 4, it may be clear that we are not really interested in the shape of W_1 as it crosses some remote point z_o. Instead, we are interested in its shape immediately after it emerges from the object F. This is especially true when W_1 is warped, since warpage changes as discussed in conjunction with Fig. 1.

How can we capture the warpage at F? We can do it with a camera, or with the human eye, or with a similar optical system.

A camera projects an image of an object. When a camera lens is focused on a specific object plane, it directs light emerging from that plane into a specific image plane (or film plane); it directs light from every object point x,y to a corresponding point x', y' in the image plane. Moreover, *the phase relationship of light that emerged from the object point is preserved for light crossing the image point.*

Figure 8 illustrates the argument. Mirror D from Fig. 4 is shown with wave trains carrying W_1 and W_2 toward the camera. The camera is adjusted such that light from plane A-A is focused on the film plane. However, light from object plane 2 lies at the same distance from the camera, and it, too is focused in the film plane. The camera cannot sense the true path of W_2, but reacts as though this light came from W_2' (everywhere equidistant from W_2), which is called the *virtual image* of W_2.

When plane wavefront W_2 crosses the image plane it is deformed into a spherical wavefront W_2''. When warped wavefront W_1 crosses the image plane, it has a new warped shape W_1'', where its original deviations from a plane reappear now as equal deviations from a sphere.

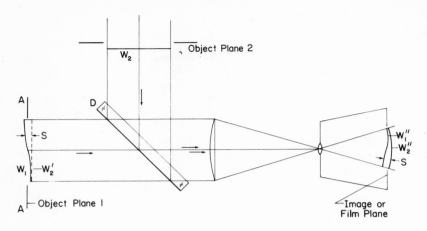

Fig. 8   Wavefront separation S in object plane is
         reproduced by camera as wave trains cross
         the image plane.

Two coherent wave trains with wavefronts W_1'' and W_2'' cross
the image plane. They generate a steady-state interference pattern
as a contour map of the separation between W_1 and W_2''. However,
this must also be the contour map of separation between W_1 and
W_2', and the objective is achieved.

INTERFEROMETRY--PRACTICAL TECHNIQUES

The four-mirror instrument sketched in Figs. 2 and 4 is
called a Mach-Zehnder Interferometer. It is invaluable for
explaining the principles of interferometry, but it is not
useful for non-destructive testing of opaque solids. Instead,
its practical application lies in tests and measurement of
transparent media. Wind-tunnel tests and tests of optical
elements are examples.

Surface Topography--Macroscopic; Fizeau Interferometer

Suppose it is important to determine the topography of a
smooth, slightly warped surface. It may be important because
the warpage was caused by erosion, by irradiation, by stress
relief, by external forces, etc.

A good method is illustrated in Fig. 9, depicting the optical setup for Fizeau interferometry. Light from the laser is
directed by the beam-splitter (R = T = 50%) toward the specimen;
the diverging beam is rendered parallel by the field lens; a

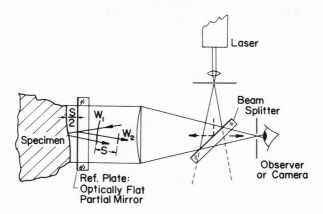

Fig. 9   Fizeau interferometer. Contour map of wavefront
          separation S formed in camera depicts topography
          of surface of specimen.

portion of the light is reflected at the mirrorized surface of
the flat reference plate, which returns wave trains with plane
wavefront W_2. Another portion penetrates the reference plate and
reflects back from the surface of the specimen; this wavefront
is warped, however, because of the specimen warpage. If $S/2$ is
the variable gap between the specimen and reference surface, the
separation between W_1 and W_2 is S. When the camera or eye is
focused on the specimen, interference produces a contour map of
S. Of course, this is also the contour map of the specimen
surface topography, where the contour interval is $\lambda/2$.

   Normal incidence is used in practice, but the incident ray
is drawn slightly oblique in Fig. 9 to distinguish it from the
reflected rays. If the specimen has modest reflectance, say
steel with R = 50%, best contrast of interference fringes would
be achieved with the reference plate coated for about 25% reflectance, 75% transmittance. Then, the beams carrying W_1
and W_2 will each have about 1/4 the incident intensity.

   Multiple reflections also occur between the reflective
surfaces, but they have substantially lower intensities and may
be neglected. Also, some light is reflected back from the
unmirrored surface of the reference plate. This surface may
be treated with an anti-reflective coating, or preferably, it
may be inclined or wedged with respect to the mirrorized surface so that this extraneous light will lie off-axis and not pass
through the opening of the aperture plate.

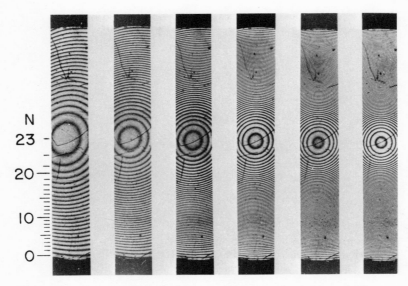

Fig. 10   Successive fringe patterns show increasing
          deformation of the surface of a disk as
          pressure on one side is increased.

In the event the specimen surface has low or uneven reflectance, it may be overcoated with a thin metallic layer. Metal
deposition by evaporation or sputtering is fairly commonplace,
resulting in coatings of only about 50 nm thickness and con-
forming to the surface topography extremely well.

An interesting example is given in Fig. 10.   The contour
maps show progressively increasing deformation of a circular disk
as pressure on one surface of the disk is increased.   The full
circular disk was observed, but only a central strip is shown here.

Displacement W normal to the disk is fringe order times contour interval, or

$$W = N(\lambda/2)$$

Displacement of the center of the disk relative to the edge is
calculated for the first case by substituting $N = 23$ and $\lambda = 632.8$ nm
(helium neon laser), giving $W = 0.00728$ mm or 0.000287 inches.

Sensitivity and Accuracy

The location of the center of a dark or bright fringe can
easily be estimated to within 1/5 the distance between adjacent

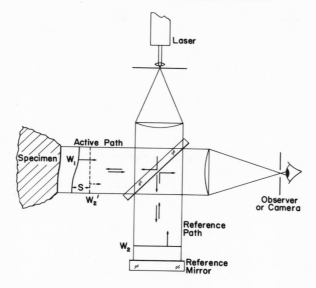

Fig. 11   Michelson interferometer.   Interference
          of coherent wave trains with wavefronts
          W_1 and W_2 forms a contour map of specimen
          topography.

fringes.   Thus, sensitivity of the measurement is 1/5 contour
interval, or 63 nm or 2.5×10^{-6} inches.

The fringe pattern provides basic data and accuracy depends
upon the technique of withdrawing the information.   If observations
of fringe orders are made at the outside boundary and at the center
of the disk, there is potential for additive errors and the accuracy can be reduced to 2 x (1/5 contour interval) or 2 x (1/5) x
($\lambda/2$), viz., accuracy of 130 nm or 5×10^{-6} inches.   On the other
hand, a plot of fringe order N, for each fringe, versus position
along the diameter of the pattern can be made.   This will
yield a faired curve that is up to an order of magnitude more
accurate than the data at any point, yielding an accuracy that
approaches 1/10 the sensitivity, or about 6 nm or 0.25×10^{-6} inches.

Michelson Interferometer

Surface topography can be mapped with the Michelson Interferometer, too, as illustrated in Fig. 11.   This scheme is more
complicated than the Fizeau system, but it is especially useful
when the reference plate cannot be located near the specimen because of size or shape limitations, environmental limitations
or otherwise.   In this case, an optically flat beam splitter
directs half the light to the specimen and the other half to a flat

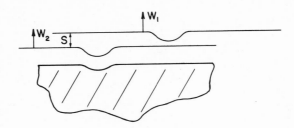

Fig. 12   Wavefront shearing microscope produces
          two mutally displaced images of the
          emergent wavefront; interference reveals
          separation S, which provides a depth
          measurement of topographical features of
          the specimen.

reference mirror.   After reflection, the two beams meet again at
the beam splitter and portions of each emerge horizontally to be
collected by the observer.   When focused on the specimen, the
interference pattern seen by the observer is the contour map between warped wavefront W_1 emerging from the specimen and plane
wavefront W_2', which is the virtual image of W_2 emerging from
the reference mirror.   By geometry it is clear that warpage of
the reflected wavefront is again twice the warpage of the specimen
surface, so the contour interval is again $\lambda/2$.   Sensitivity and
accuracy are the same as in Fizeau interferometry.

      The Michelson interferometer is important historically be-
cause lengths of the active and reference paths can be made nearly
equal, thus circumventing the need for highly coherent light with
long wave trains.

Surface Topography - Microscopic

      It is often important to inspect surface topography on a
microscopic scale, for example in fracture studies, surface
finish studies, and scratch depth studies.   Metallographic microscopes, i.e., microscopes equipped for through-the-lens illumination, may be fitted with various commercial interferometer
nosepieces.   One popular type is a Michelson Interferometer
attachment, in which a small 45° beam splitter and a reference
mirror are fitted between the microscope objective lens and the
specimen.   Since the active path and reference path are nearly
equal in length, laser light is not required; instead, the fairly
monochromatic light of a filtered mercury vapor lamp, or similar
source, can be used.

As a simple, non-commercial alternative, two-beam Fizeau fringes can be generated in a microscope if a thin, partially mirrorized glass plate is laid over the surface of the specimen. A microscope cover slip as thin as 0.1 mm, coated for 25-50% reflectance, is suitable. Laser light, or a narrow spectrum laboratory light source should be used.

Another simple interferometric attachment is suitable for measurement of scratch profiles, cleavage steps, or similar isolated details. Let the specimen illustrated in Fig. 12 have a fairly smooth reflective surface with a narrow groove or blemish. If a plane wavefront is incident on the surface the reflected wavefront will have the form depicted by W_2. The microscope attachment utilizes a doubly-refracting crystal to generate a second wavefront W_1, which has the same shape as W_2 but is laterally displaced as shown. Interference then provides the contour map of separation S between the wavefronts, which closely approximates the depth contours of the groove or blemish. The method is called *wavefront shearing interferometry*, since the wavefront is divided into two equal parts and one of these is displaced, or sheared laterally, with respect to the other.

Surface Topography - Multiple Beam Interference

In a quick departure from the two-beam interference theme, multiple-beam interference should be mentioned. The optical arrangement is identical to that of Fig. 9, except that the specimen surface and reference plate are highly reflective; the latter may have 85-95% reflectance and 15-5% transmittance. Multiple reflections experienced by a ray are illustrated in Fig. 13. Intensities of successive emergent contributions drop off gradually but many contributions are effective in establishing the resultant intensity. Actually, normal incidence is used. The multiply-reflected rays travel the same path between the mirrors, generating a consistent wavefront lag of S in each double traversal, and all the rays emerge along a common line.

The resultant output intensity varies with S as shown in Fig. 13b. Thus, the multiple-beam interference pattern is again a contour map of the gap between specimen and reference plate and contour interval is again $\lambda/2$, but the fringes are now narrow dark lines on a bright background. An advantage lies in the ability to determine the center of the dark fringe with greater accuracy than possible with the broader two-beam fringes. However, its practical application is limited to surfaces with more modest warpage.

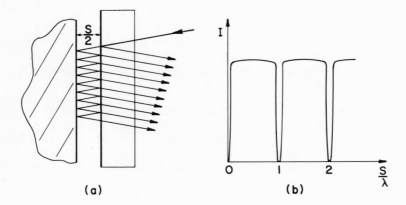

(a)   (b)

Fig. 13 Multiple beam interferometry. (a) Emergent
         components of input beam travel systematically
         varying path lengths; (b) these components
         combine to form sharpened interference fringes,
         with narrow dark bands on a bright background.

Thermal Expansion - Relative

Interferometry is exceptionally well suited for measurements
of coefficient of thermal expansion (CTE). Figure 14 illustrates
one method, using Fizeau interferometry. The specimen to be
measured is installed between the fully-reflective and partially
reflective mirrors, together with two reference bars of known CTE.
The assembly is installed in an oven or furnace and viewed with
the external apparatus shown in Fig. 9.

If the specimen and reference bars are originally machined
to identical lengths, the mirrors will be exactly parallel and
the contour map of their separation would show a null field, i.e.,
constant fringe order N throughtout the field at view. Then, as
temperature is increased and the specimen expands relative to
the reference bars, the mirrors become increasingly non-parallel
and a pattern of uniformly spaced fringes (Fig. 14b) appears.
CTE is calculated by

$$CTE = \frac{\Delta L/L}{T-T_o} + CTE_{ref}$$

where

$$\Delta L = \frac{dN}{dx} (h) \left(\frac{\lambda}{2}\right)$$

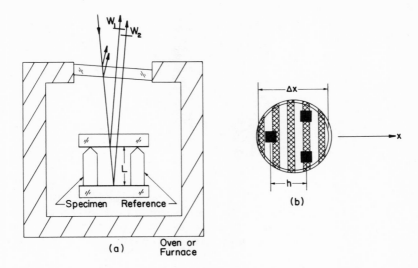

Fig. 14   Fizeau interferometry for thermal expansion
of specimen relative to reference bar.

Fringe order gradient dN/dx may be determined conveniently by
counting the change of fringe order ΔN in a predetermined length
Δx (a circle of diameter Δx may be engraved on one of the mirrors).
λ/2 is the contour interval of the interference pattern, and of
course it must be expressed in the same units as other length
dimensions in the calculation.

Sensitivity is proportional to specimen length L, and with
ordinary laboratory lasers L may be as much as a few centimeters.*
However, when L is a few centimeters and temperature range is
substantial, thermal air currents between the mirrors cause the
fringes to move, or dance, enough to make measurements difficult
or impossible. The best solution appears to be vacuum, either
by surrounding the apparatus with a small vacuum chamber placed
in the oven or by evacuating the oven itself. Incidentally, windows in the vacuum chamber and oven need not be optical quality,
since any wavefront distortions they introduce effects W_1 and W_2
equally and thus does not change the fringe pattern. The windows

---

*What is the limitation? Ordinary laboratory lasers have much
greater coherence lengths, but drift of cavity length and longitudinal mode shifting introduce small changes of wavelength with
time. Since the difference of OPL of interfering beams is large
in this experiment, viz. 2L, the number of wavelengths that fit
into this distance is highly sensitive to small changes of the
wavelength. Stabilized lasers are available with wavelength
stability improved by several orders of magnitude.

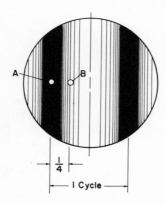

Fig. 15 Photoelectric cells A and B are located 1/4
fringe apart in the image plane. Simultaneous
monitoring as fringe orders change reveals
whether fringe order is increasing or decreasing.

should be slightly inclined with respect to the interferometer
mirrors so that light reflected from their surfaces will have
different directions and will be blocked by the aperture plate
in front of the camera.

The major advantage of this relative expansion method is that
the relevent fringe count can be made at any fixed temperature;
continuous monitoring and fringe counting during the period of
temperature change is not required.

Thermal Expansion - Absolute (Quadrature Signals)

The experimental setup for absolute measurement of CTE can
be the same as that in Fig. 14, except all three spacers are made
of the specimen material. Then, the mirrors remain parallel as
the specimens expand and output light intensity changes cyclically
from maximum to minimum to maximum, etc. It is necessary to
monitor the intensity changes continuously to determine the change
of fringe order and the change of specimen length.

It is possible to monitor a test visually, but if numerous
tests are required it is more practical to monitor the light output with a photoelectric cell and strip-chart recorder. Let the
photoelectric cell be installed at a point in the film plane of
the camera. It will respond to changes of intensity by generating
a proportional change of electrical current, which may be amplified
and recorded automatically as a graph. Alternatively, the
cycles of output current may be counted electronically.

This arrangement is satisfactory if the temperature and fringe
order are changing monotonically, constantly increasing or

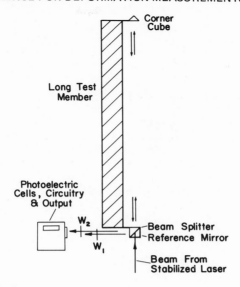

Fig. 16 Michelson interferometer with automatic
fringe counting accessories measures small
changes of large lengths.

decreasing. It is difficult, however, to assure against a tem-
porary reversal of temperature and fringe order change, and if a
reversal occurs, ambiguities arise in the fringe order count. The
problem is solved by utilizing two photoelectric cells as indicated in Fig. 15. In this case, one of the three specimens is
made slightly shorter, so that a gradient of at least one fringe
order appears in the fringe pattern in the film plane. Photocells
A and B are then located at points 1/4 fringe apart, as shown.
The electrical output signals will exhibit the same variation, but
90° out of phase. These are called quadrature signals and they can
be circuited to sense whether the cyclic function is increasing or
decreasing and add or subtract accordingly. Automatic fringe
counting is accomplished without ambiguity for any variation of
fringe order.

The least complex quadrature fringe counters give a least
count of 1/4 fringe, and since the contour interval is $\lambda/2$ per
fringe, the least count represents a change of length of $\lambda/8$. For
helium-neon laser light, this is about 80 nm or 3×10^{-6} inches
per count. More sophisticated systems are capable of greater
sensitivity by one to two orders of magnitude.

Expansion and Displacement Over Long Distances

Sometimes it is necessary to ascertain thermal expansion
of very large members, perhaps several meters long. One example

is a space telescope, where members connecting optical elements
must have virtually zero CTE. A method utilizing a Michelson
interferometer system is illustrated schematically in Fig. 16.
A beam-splitter cube is fixed to one end of the member and a
corner cube to the other end. A portion of the incoming laser
light reflects at the beam-splitter, travels a short fixed path
length to the reference mirror (which is a mirrorized face of
the beam-splitter cube) and emerges horizontally with wavefront
W_2. Another portion is transmitted at the beam-splitter, travels
to the corner-cube and back, and emerges from the beam-splitter as
W_1. (A corner cube acts like a mirror, but it is easy to align
since it returns light along its original path regardless of its
orientation.) W_1 and W_2 form an interference pattern with a
gradient adjusted to about one fringe. Quadrature signals are
sensed and the electronic package shows changes of length between
the beam-splitter and corner cube.

For systems such as these, where the path lengths traveled
by W_1 and W_2 are grossly different, small changes in wavelength
would be devastating. Specially stabilized lasers are essential.

An especially important application of precision measurements
over long distances is in the manufacture of large machined
parts. Special machine tools are fitted with automatic fringe
counting interferometers which measure the position of the tool
carriage with respect to any preset starting position. Carriage
displacements may be 10 meters or more. Special machines are
fitted with interferometers for precision measurements along
x-y axes, and even x-y-z axes.

Commentary

Interferometry is a powerful tool for NDT measurements of
solids. In most instances it provides precision measurements of
displacements in the direction of observation, e.g., out-of-plane
displacements in topography studies. For surface topography,
classical interferometry is limited to study of relatively smooth
surfaces. The surfaces must be regular enough that the fringes
are clearly visible with ordinary means of viewing, e.g., by
camera for macroscopic surfaces or by optical microscopes for
small fields of view. For rougher surfaces the interference
fringes will be too close together and another method called
holographic interferometry is required.

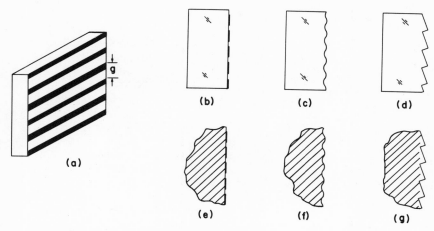

Fig. 17 (a) A diffraction grating is any cyclical
        irregularity of a surface. (b), (c) and (d)
        are transmission gratings, where (b) is a
        bar-and-space amplitude grating, (c) is
        a symmetrical phase grating and (d) is a
        blazed phase grating; (e), (f) and (g) are
        corresponding reflection gratings.

MORE FACTS AND FUNDAMENTALS

Diffraction Gratings

Holography and moiré depend upon diffraction of light as well
as interference, and their explanation requires an understanding
of diffraction by gratings. The fundamentals of grating behavior
will be discussed here in sufficient depth for both subjects.

Diffraction gratings are illustrated in Fig. 17, where (a)
indicates that a grating is a surface with regularly spaced bars
or furrows. In the case of transmission gratings, the incident
light and diffracted light appear on opposite sides of the
diffracting surface; with reflection gratings, they are on the
same side and the substrate may be opaque. *Amplitude gratings*
consist of opaque bars and transparent spaces, or else reflective
bars and non-reflective spaces. They are so named because the bars
periodically alter the amplitude of the incident light, where the
period or *pitch* is g. *Phase gratings* have furrowed or
corrugated surfaces, with either symmetrical or non-symmetrical
furrow profile. These alter the phase of the incident light in a
regular, repetitive way.

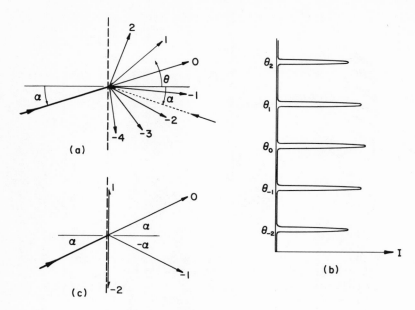

Fig. 18 A grating redistributes the intensity of an incident beam into preferred directions, shown as diffraction orders in (a) and the corresponding intensity distribution for an unblazed grating in (b). (c) illustrates the special case of symmetrical diffraction orders.

Frequency f of a grating is the number of bars or furrows per unit length, usually expressed in "lines" per unit length (meaning repetitions or cycles per unit length). Frequency and pitch, g, are related by

$$f = \frac{1}{g} \qquad (11)$$

Our interest lies in relatively high frequency gratings from perhaps 100 lines per mm (ℓ/mm) to 2 or 3 thousand ℓ/mm.

A grating divides every incident wave train into a multiplicity of wave trains of smaller intensities; and it causes these wave trains to emerge in certain preferred directions. This is indicated in Fig. 18, in which the dashed line may represent either an amplitude or a phase grating. When a parallel beam is incident at angle α from the left, the grating divides it into a series of beams which emerge at preferred angles: ---, $\theta_{-1}, \theta_0, \theta_1, \theta_2$ ---. These beams are called *diffraction orders* and are numbered in sequence beginning with the zero order, which is an

extension of the incident beam; counter-clockwise diffractions
are considered positive.

Reflection gratings diffract light in just the same way, except the incident light is the mirror image of that for a transmission grating, as shown by the dotted ray in Fig. 18(a). In
reflection, the 0th-order diffraction is the mirror reflection
of the incident beam.

The *grating equation* defines the angles of diffraction, viz.,

$$\sin \theta_m = m\lambda f + \sin\alpha \tag{12}$$

where m denotes the diffraction order of the ray or beam under
consideration. The same units of length must be used in λ and f.

Much of our interest will focus on the special case of Fig. 18c
where the 0th diffraction order and its neighbor, the -1 order, is
symmetrical with respect to the grating normal, i.e., with
respect to $\theta = 0$. Since $\theta_0 = \alpha$, the direction of its symmetrical
neighbor must be $\theta_{-1} = -\alpha$, and for this condition Eq. 12 reduces
to

$$\sin\alpha = \frac{\lambda}{2} f \tag{13}$$

This defines the angle of incidence α to achieve the special condition of symmetry.

Angle of diffraction θ_m cannot exceed +90°. It is clear from
Eq. 12 that the number of diffraction orders that emerge within
the range ±90° is a very large number when f is small (i.e.,
for a coarse grating) and conversely the maximum value of m is
small when f is large. The angle between neighboring diffraction
orders is small for a coarse grating and it is large for a fine
grating.

For the case of symmetrical 0th and 1st order diffractions
(Eq. 13), f approaches its maximum value as α approaches 90°
and $\sin\alpha$ approaches unity. In the limit, the maximum value of f
is

$$f_{max} = 2\left(\frac{1}{\lambda}\right) \tag{14}$$

In red light, this corresponds to about 3200 ℓ/mm (80,000 ℓ/in)
and in blue-green to about 4000 ℓ/mm (100,000 ℓ/in). The corresponding pitch of the grating is $\lambda/2$, or half a wavelength between
neighboring lines.

Another interesting limiting case is sketched in Fig. 18c.
Under what conditions do only two symmetrical diffractions emerge

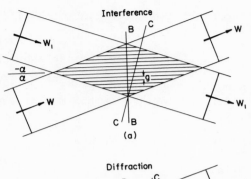

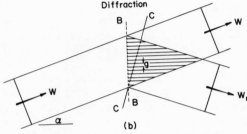

Fig. 19   Interference of two coherent beams (a) forms
          a virtual grating of frequency f, while (b)
          diffraction from a real grating of frequency
          f produces the same two beams and the same
          virtual grating.

from the grating? This occurs when angle α is as shown, or any
greater angle up to 90°. As shown, the angle of the +1 diffraction order is 90°, or θ_1 = 90°. By substituting Eq. 13 (for
symmetry), m = 1 and θ_1 = 90° into Eq. 12, we find

$$f = \frac{2}{3}(\frac{1}{\lambda})$$ (15)

and the corresponding pitch is g = $(3/2)\lambda$. Thus, the conditions
under which two, and only two, diffraction orders emerge from the
grating are known, viz.,

$$2/\lambda > f \geq 2/3\lambda$$

or   (16)

$$\lambda/2 < g \leq 3\lambda/2$$

Equivalency, Virtual Gratings

     Take notice of Eq. 9. It relates the half-angle α between
two coherent beams and the pitch g of the interference pattern

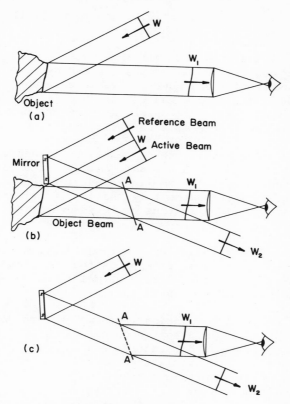

Fig. 20   In holography, photographic plate A–A captures
          the interference pattern formed by the object
          beam and reference beam. When developed, the
          photographic plate becomes a grating (here
          called a hologram); in (c) the reference beam
          passes unmodified in the zeroth order of the
          grating, but in the first order the angle of
          diffraction of every ray is exactly right to
          reconstruct the wavefronts of the object beam.
          Thus, the observer sees exactly the same view
          in (a) and (c).

thus formed. If the frequency f of the interference pattern is
$f = 1/g$, the equation becomes identical to Eq. 13!

     Remember the photographic plate B–B exposed to the interference
pattern in Fig. 5. The exposed and developed plate is a pattern of
parallel, equally spaced bars governed by Eq. 9; it is a diffraction
grating of pitch g and frequency f. If this diffraction grating
is illuminated at a suitable angle α, it will generate 0th and

1st order beams governed by Eq. 13, equal to those that originally formed the grating.

Figure 19 illustrates the relationship. In the volume of space where two coherent beams coexist, a steady-state formation of walls of constructive and destructive interference is generated. On the other hand, a real grating of the same pitch g creates two coherent beams that coexist in space and they create the same walls of interference in their zone of intersection.

This three-dimensional zone of regularly spaced walls of interference will be called a *virtual grating* since this light possesses the same properties as light emerging from a real grating. The significant concept is that a virtual grating can be formed either by a real grating or by intersection of two coherent beams.

While the equivalence relationship was derived for plane B-B in Fig. 19, we can prove that it applies for any plane C-C. Thus, symmetry with respect to the grating normal is not a necessary condition. The generalized conclusion is:

> If a diffraction grating is formed by exposing a
> photographic plate to two intersecting coherent
> beams, then that diffraction grating will re-
> produce the two original beams (in its 0th and
> 1st orders) when illuminated by either one of
> the original beams.

This concept is the foundation of holography.

HOLOGRAPHY

Holography is a unique means of capturing the wavefront shapes emerging from solid objects and storing them for future interrogation. With our background, the concepts of holography are easily understood.

Consider Fig. 20a, in which an object with a smooth warped surface is illuminated by a parallel beam of coherent light. Light reflected by the object is collected by the field lens and viewed by an observer or camera. The observer sees the object in its true perspective, including all visible details of shape the object may have.

Let a mirror be installed next to the object as shown in (b) so that coherent light of the reference beam crosses the beam emerging from the object. A virtual grating of walls of construc-tive and destructive interference is formed in the crossing zone.

The virtual grating does not have perfectly plane and equally
spaced walls, however, because the object beam has warped wavefronts.
Let a photographic plate be installed in the virtual grating at
A-A. It is exposed to light where it crosses walls of constructive interference and it receives no exposure (no light energy)
where it crosses adjacent walls of destructive interference. Upon
development,the photographic plate becomes a real grating; again,
the lines are not perfectly straight and equally spaced. Instead,
they are irregular, in conformity with the shape of wavefront
W_1 that was used to form the grating.

Next, install the grating into its original position A-A and
remove the object, as sketched in (c). Light from the reference
beam passes through the grating undisturbed in the 0th order,
since θ_o is not effected by the frequency of the grating. However,
at every point in the grating, light of the +1 diffraction order
is diffracted in exactly the direction of the original object
beam, so that wave trains with wavefronts W_1 are generated to the
right of the grating. Of course, this is a consequence of the
equivalency relation between interference and diffraction. Light
collected by the observer has exactly the same content as it had
in Fig. 20a!

The observer does not see the object itself in (C), but he
sees a virtual image of the object. It conveys to the observer
the same information as he received in (a), including any depth
or 3-dimensional information. The grating that stores this
information is an interference fringe pattern, but in this
technique it is called a *hologram.*

Perhaps it is apparent that the reference beam need not be
parallel, as long as the same reference beam is used in reconstruction as was used to expose the hologram. For convenience, a
diverging beam is usually used, as sketched in Fig. 21.

Perhaps it is also apparent that the wavefront emerging from
the object need not be smooth or continuous. The hologram may be
formed with light of any wavefront shape, and light in its first
diffraction order will reproduce a virtual image of that shape.
Let the object have a rough or matte surface, such as obtained
with a flat (non-glossy) paint. Diffuse reflection will result;
every point on the object will reflect an incoming ray as a
spherical wave train, sending light out in all directions, as
indicated in Fig. 21. The envelope of these spherical wavelets
is the object wavefront W_1, which in general will have a very
irregular complicated shape relating to the shape of the object.

Since light emerges from the matte object in all directions,
it can be viewed from different stations, for example from B or C.

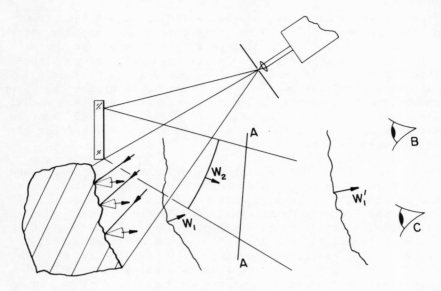

Fig. 21 Holography with a diffuse or matte object.

Similarly, in the reconstruction step, light carrying a virtual image of the object can be viewed at B or C. The boundary of the hologram is like a window, and any part of the object that can be seen through a window of that size becomes visible in the reconstructed image.

Since the object wavefront can be so irregular, the hologram must contain lines (interference fringes) that are very closely spaced. Extremely high resolution of photographic plates are required. However, photographic materials are available that will resolve 4000 ℓ/mm or more, meeting the theoretical limit of grating frequency given by Eq. 14.

HOLOGRAPHIC INTERFEROMETRY

Consider Fig. 20b again, but this time let the developed hologram be reinserted in position A-A. Light from the reference beam reaches the observer through first order diffraction and appears as a virtual image of the object. Simultaneously, light from the active beam is reflected by the real object and passes through the zero diffraction order to reach the observer. The observer sees the real object and the identical virtual image, and they appear as one image without distinction.

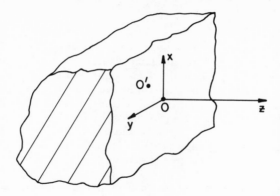

Fig. 22   Displacement of point 0 in a surface to 0'
           has components U, V, W in the x,y,z directions,
           respectively.

     Now, let a small change occur in the surface of the object,
perhaps caused by external loading, by thermal effects, or
otherwise. The wavefronts comprising the light coming from the
object are slightly different than before, but the wavefronts
from the virtual image, reconstructed from light diffracted by
the hologram, remain unchanged. When the light reaches the retina
of the observer, or the film plane of the camera, the wavefronts
are not identical and they combine in constructive and destructive interference. The result is an image of the object
crossed with interference fringes. The fringes form a contour
map of displacements on the surface of the object--fringes
that reveal the displacements accrued after the hologram was
made.

     The impact of holographic interferometry lies in its
ability to make subwavelength measurements on irregular bodies.
With respect to classical interferometry, the crucial difference
is this:

          whereas in classical interferometry the observer
          must resolve interference fringes associated with
          the topography of a body, in holographic interfer-
          ometry the observer is called upon to resolve only
          those fringes associated with its change
          of topography.

Thus, the complexity of topography is of no consequence in
holographic interferometry, and rough or deeply convoluted

surfaces remain applicable.

But what displacements do the fringes represent? In Fig. 22, let 0 represent a point in the surface, which after deformation is displaced to point 0'. Let U, V, W represent the displacements of 0 in the x, y, z directions, respectively. If the apparatus is arranged for direction of illumination and observation both normal to the surface--by use of beam-splitters--then the contour map has the same meaning as in classical interferometry. It represents the out-of-plane displacements, W, and the sensitivity or contour interval is again $\lambda/2$ per fringe.

If direction of object illumination and/or direction of observation is different from normal, the contour map becomes a function of the three displacements at every point, U, V and W. It is possible to determine the complete U, V, W displacement field by means of three independent fringe patterns, e.g., holographic interference fringe patterns obtained with three different directions of illumination. Here, the basic data is the fringe order at (any) point 0 from each pattern and the respective directions of illumination and observation. The method has a sound foundation in theory, but accurate analysis is difficult.

The method described above is called *real-time* holographic interferometry, since one set of wavefronts comprising the interference pattern emanate from the object during the actual time the pattern is being observed. An alternative is to let both sets of wavefronts--before and after displacement of the object--emanate from the hologram. This is called *double-exposure* holographic interferometry. Referring again to Fig. 20b, the photographic plate is exposed to the virtual grating formed when the body is in its original position. Then the body is deformed or otherwise displaced and a second exposure is made on the same photographic plate. When developed, the holographic plate contains two superimposed diffraction gratings; in Fig. 20c, they produce wave trains with two different wavefronts, W_1 and W_1' emerging in the +1 diffraction order. Of course, these are reproductions of the wavefronts that emanated from the object in its original and displaced positions, and the resultant interference pattern is a contour map of the displacements. The result is the same as in the real-time technique, but all the information is stored in the hologram for future retrieval.

APPLICATIONS

The most significant application of holographic interferometry in NDT is flaw detection. As an example, consider a panel made up

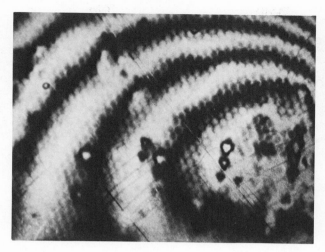

Fig. 23   Holographic interferometry used as NDT of
honeycomb panel.   Internal defects appear as
anomalies in the otherwise regular pattern.
Courtesy of Gene E. Maddux, A. F. Flight
Dynamics Laboratory.

of a honeycomb core laminated between two continuous surface skins.
If stresses in the top surfaces are developed by changing the
external pressure or by quickly heating the surface, small displacements of the skin could be observed by holographic interferometry.   An example is shown in Fig. 23 which is a double-
exposure hologram of such a panel.   The panel was heated between exposures and the large fringe loops show the general W-
displacement field caused by the heating.   Small hexagonal features are visible, especially in the transition from dark to
bright fringes; these are localized zones of slightly smaller
displacements where the honeycomb core elements were bonded to
the skin and restrained its displacement.   Internal defects show
up as anomalies in the otherwise regular fringe pattern.   They
are seen as dark spots on bright fringes and bright spots on dark
fringes; toward the left side of the central loop, three defects
are seen in a vertical row.   These defects all represent honeycomb
cells that are weakened by corrosive attack on the metallic
honeycomb material.   The panel skin is not properly reinforced by
these cells, but instead it is allowed to bulge locally when
heated and appear as localized excess W-displacements.

A similar example is flaw detection in pneumatic tires.   Let
a double exposure hologram be made of a tire, but with a small
change of internal pressure between exposures.   The image of the
tire is crossed with a smooth array of interference fringes,
depicting the displacements caused by the change of pressure.   In

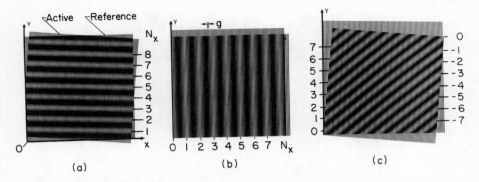

Fig. 24 Mechanical interference by obstruction of light
         for two coarse bar-and-space gratings. Displace-
         ment of active grating is (a) pure rotation,
         (b) pure extension, and (c) rotation plus com-
         pression.

regions weakened by delamination between plies, or torn belts,
or other defects, the deformation will be excessive, and such
zones will again appear as local concentrations of displacement
fringes.

MOIRÉ INTERFEROMETRY

Background, Mechanical Interference

     Alas! The fundamentals needed to understand moiré inter-
ferometry have already been covered. But still, discussion of
a closely related subject, mechanical interference, or coarse
moiré, would be useful.

     Whereas classical and holographic interferometry are most
effective for measuring W, the out-of-plane displacement, the
moiré methods to be described here are employed to measure U
and V, the in-plane displacements.

     Coarse moiré is illustrated in Fig. 24. It is produced
simply by the superposition of two coarse amplitude gratings,
with frequencies in the range of about 1 to 40 ℓ/mm. Let one
be called an *active* grating and the other a *reference* grating.
In (a), they have equal frequency, but the active grating is
given a rigid body rotation about point 0. Zones of light
and dark bands--again called fringes--are seen where the grating
lines are in registration and a half-cycle out of registration,
respectively. Inspection of Fig. 24 shows that for the light

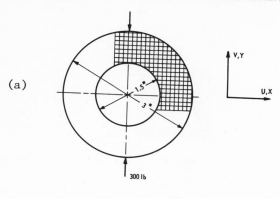

(a)

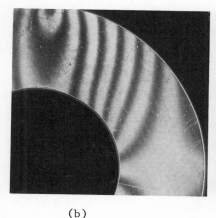

(b)

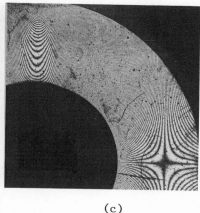

(c)

Figure 25   (a)   Ring with 20ℓ/mm cross-line grating.
            (b)   V-displacement field by coarse moiré
            method.   (c)   Identical V-displacement
            field with moiré fringe multiplication by
            a factor of 20.

fringe above 0 all the lines in the active grating are shifted
one pitch, g, to the right; on the next light fringe above, all
the lines are shifted two pitches in the x-direction; etc.

It becomes clear that moiré fringes are the
locus of points of constant displacement,
specifically the in-plane displacement component in the direction perpendicular to
the lines of the reference grating.

Moiré fringe order, N, denotes the number of cycles of intensity
fluctuation (light to dark to light, etc.) experience at any
point as the displacement changed from zero to its final value.

Its relation to displacement is

$$U = gN_x$$

$$V = gN_y$$

(17)

where

g is the pitch of the reference grating,

U and V are displacement components in the x- and y-directions, respectively,

N_x and N_y are fringe orders when lines on the reference grating are perpendicular to the x- and y-axes, respectively.

Fringes of pure rotation (a) bisect the angle between normals to the grating lines; when the distance between neighboring fringes is very large compared to pitch g, fringes of rotation N_x may be considered parallel to the x-axis.

In (b), the active grating was stretched relative to the reference grating and the resultant fringes are perpendicular to the x-axis. In (c), the active grating was compressed and rotated, both, and fringes are inclined to the axes. Equations 17 apply for every case.

Coarse moiré may also be used for measurements in non-homogeneous displacement fields, as illustrated in Fig. 25. A coarse grating of 20 ℓ/mm was printed onto the surface of a circular ring as indicated in (a). In this case the grating was a cross-line grating* and it could be interrogated by superimposing a uniaxial grating in either direction. The pattern shown in (b) was obtained by applying external loads to the ring, thus deforming it and the active grating, and then superimposing a 20 ℓ/mm reference grating with lines perpendicular to the y-axis. The resultant moiré pattern is a contour map of the V-displacement field in the deformed ring.

For a structural part, the U- and V-displacement fields are important in their own right. They are valuable too because they can give strain and stress fields.

For strains,

---

*A cross-line grating has bars or furrows in two perpendicular directions and diffracts light in both the xz and yz planes, simultaneously.

$$\varepsilon_x = \frac{\partial u}{\partial x}; \quad \varepsilon_y = \frac{\partial v}{\partial y}; \quad \gamma_{xy} = \frac{\partial u}{\partial y} + \frac{\partial v}{\partial x} \tag{18}$$

where ε_x and ε_y are normal strains in the x and y directions, respectively, and γ_{xy} is shear strain in the xy plane. Then, stresses may be determined using the stress-strain relations for the structural material. Since strains and associated displacements are usually very small, most practical applications require the high sensitivity obtained when g is very small (Eq. 17), i.e., gratings of fine pitch and high frequency are required.

Coarse moiré is usually explained as mechanical interference, i.e., the mechanical obstruction of light by the opaque bars of the gratings. The explanation is entirely satisfactory for coarse amplitude gratings in intimate contact. However, as the pitch of the grating becomes smaller, or as the gap between the active and reference gratings increases, or both, the simple means of observation and the simple mechanical explanation become inadequate.

Instead, the techniques of moiré interferometry must be used. These extend the domain of moiré to approach the theoretical limit given by Eq. 14--to thousands of lines per millimeter. Moiré experiments with grating frequencies as close as 96% of the theoretical maximum have been successful.

For coarse moiré, the active grating may be either a transmission or reflection type. The same is true for moiré interferometry. Most applications of NDT will involve opaque specimens and the reflection approach would be required. However, the explanation is a bit less encumbered for transmission systems and those will be described first.

Moiré Interferometry, Transmission

The original technique is illustrated in Fig. 26a. It was characterized in detail in a 1956 monograph by Guild.[1] We will restrict our attention to the case of symmetrical double-order dominance, i.e., where the zeroth and first diffraction orders emerging from the reference grating are much more intense than any other orders, and where these two dominant orders are symmetrically oriented with respect to the normal to the reference grating. The latter condition is accomplished when angle of incidence α is given by Eq. 13; it is important because this symmetry makes the moiré interference pattern insensitive to out-of-plane displacements. Double-order dominance is automatically achieved with fine gratings when grating frequency is in the range given by Eq. 16, or frequencies from about 1100 to 4000 ℓ/mm (27,000 to 100,000 ℓ/in). It is also achieved when the reference grating is produced

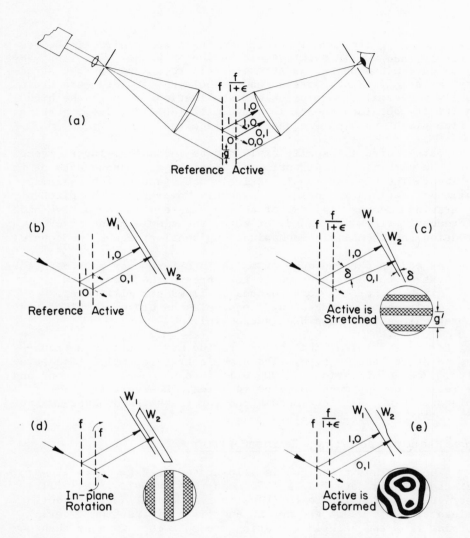

Fig. 26 (a) Moiré interferometry with symmetrical illumination
and viewing directions. Wavefronts and corresponding
interference (or moiré) pattern for (b) null field,
(c) pure extension, (d) in-plane rotation and (e) general
deformation.

by holographic techniques to produce a so-called "thick hologram", which functions by Bragg reflection. Double-order dominance can also be achieved for lower frequencies by commercial blazed gratings selected by the criteria of Post.[2]

The active grating has nearly the same frequency, viz. $f/(1+\epsilon)$ where ϵ is small compared to unity. It causes nearly the same angular deviation of the first-order diffraction.

The active grating produces diffraction for each incident beam. The numbers attached to the rays are the diffraction orders experienced at the reference and active gratings, respectively. Rays (and corresponding beams) labeled 0,1 and 1,0 emerge in approximately equal directions and they are received simultaneously by the eye or camera for observation.

When $\epsilon = 0$, conditions detailed in Fig. 26b prevail. The direction of beam 1,0 is unaffected by the active grating since it passes through it in its zeroth order. Beam 0,1 experiences the identical diffraction at the active grating as beam 1 experienced at the reference grating. Consequently, beams 0,1 and 1,0 are exactly parallel, their wavefronts W_1 and W_2 are exactly parallel, and optical interference produces a field of uniform intensity. Of course, this is exactly the same results as developed by mechanical interference, with two coarse gratings of equal pitch and parallel orientation.

When $\epsilon \neq 0$, Fig. 26c applies. Beams 1,0 is unchanged but beam 0,1 is diffracted through a slightly smaller angle since the active grating is slightly coarser than the reference. The beams and corresponding wavefronts emerge with a relative angle δ creating a pattern of horizontal interference fringes as sketched in the insert. Of course, this is exactly the same result as developed by mechanical interference with two coarse gratings of unequal pitch.

As always, the fringe pattern is the contour map of the gap between wavefronts W_1 and W_2, as specified by Eq. 10. The separation, g', between adjacent fringes depends upon angle δ and the relationship is given by Eq. 9 if $\delta/2$ is substituted for α and g' replaces g.

In Fig. 26d, ϵ is zero, but the active grating is given a small inplane rotation. Again, W_1 is not changed, but ray 0,1 is rotated out of the plane of the diagram, such that the arrowhead falls below the plane of the page. The result is inclination of wavefront W_2 relative to W_1. The contour map of their separation is given by vertical fringes as shown, which is the same result of in-plane rotation as in the case of coarse mechanical interference.

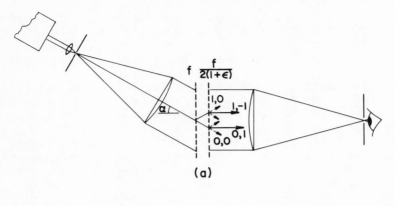

$$f \quad \frac{f}{2(1+\epsilon)}$$

(a)

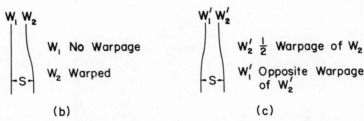

W₁ No Warpage

W₂ Warped

(b)

W_1' $\frac{1}{2}$ Warpage of W_2

W_1' Opposite Warpage of W_2'

(c)

Fig. 27 (a) Moiré interferometry with fringe multi-
        plication by 2.   (b) Wavefronts for case of
        Fig. 26e; (c) wavefronts for this case, with
        identical deformation of active grating. Re-
        sultant contour maps from (b) and (c) are
        identical.

Finally, Fig. 26e represents the general case in which ε
varies from point-to-point in the active grating, just as if the
grating was attached to the surface of a specimen and the specimen
was deformed nonhomogeneously by external forces. The frequency
and in-plane rotation vary across the active grating, causing 0,1
rays to emerge from different points in various different directions
and generate warped wavefront W_2. Now interference depicting the
gap between W_1 and W_2 yields an irregular contour map as sketched.
It is a contour map of the V-displacement field for the deformed
grating, just as in the case of mechanical interference with a
correspondingly deformed coarse grating.

Accordingly, it is clear that moiré interferometry responds to
displacements of the active grating in just the same way as coarse
moiré or mechanical interference. In fact, the relationships are
identical and Eq. 17 applies equally to both. The fringe order at
any point in the field is proportional to the in-plane displacement

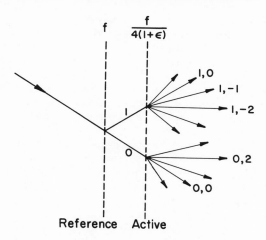

Fig. 28   Diffraction sequences when active grating is
          nominally 4 times coarser than reference
          grating.

to that point, where the constant of proportionality is the pitch
g (or reciprocal of frequency, 1/f) of the reference grating. The
important difference is that g may be exceedingly small with
moiré interferometry, allowing displacement measurements of very
high sensitivity.

Moiré Fringe Multiplication

     One shortcoming of the method of Fig. 26a is that the active
grating is viewed at oblique incidence. Thus, the view is fore-
shortened and distorted. In addition, it is not at a fixed distance from the observer or camera, and focusing might be trouble-
some for some situations. An alternative scheme shown in Fig. 27
circumvents these difficulties.

     In this case, the reference grating is unchanged, but the
active grating is everywhere half the frequency of the previous case.
When $\varepsilon = 0$, beams 0,1 and 1,-1 are diffracted by the active grating
through angles $\pm\alpha$ respectively; they emerge parallel to the axis of
symmetry, producing parallel wavefronts and an interference pattern
of uniform intensity throughout the field. When $\varepsilon \neq 0$, beams 0,1
and 1,-1 emerge with angles $\pm\delta/2$ with respect to the horizontal axis
of symmetry and produce the same pattern as in Fig. 26c. Even for
the general case, rays 0,1 and 1,-1 emerging from any point of the
active grating will always be symmetrical.

     Figure 27b and c compare wavefront warpages for the general case

of active grating deformation. In (b), representing the method of
Fig. 26, wavefront separation S is always between a plane wavefront
and the warped wavefront generated by the active grating. In (c),
representing fringe multiplication, S is everywhere the same as before, but the warpages of W_1' and W_2' are half that of W_2 and they
are everywhere equal and opposite, or symmetrical with each other.
The resultant moiré pattern--the contour map of S--is the same as
before, but the active grating is viewed at normal incidence.

The method can be carried further to utilize coarser active
gratings. In Fig. 28, frequency of the reference grating remains f,
but that of the active grating is nominally 1/4 of f. Twice as many
diffraction orders emerge from the system, and beams 0,2 and 1,-2
emerge with essentially horizontal directions. In fact, their directions are identical to beams 0,1 and 1,-1 of Fig. 27 (as can readily
be proved by Eq. 12 using frequency of the active grating in each case
so the wavefronts and the resultant interference patterns are identical.

To generalize, nominal frequency of the active grating may be
reduced to f/β, where β is any even integer. Beams emerging essentially horizontally, viz. beams $0,\beta/2$ and $1,-\beta/2$, are collected and
observed. Again, wavefront warpage is identical to W_1' and W_2'
(Fig. 27) and the same interference or moiré pattern is produced.

Curiously enough, the fringe patterns are independent of β.
Sensitivity of displacement measurements depends upon the frequency
or pitch of the reference grating (Eq. 17) and a much coarser active
grating may be employed. But why is the method called moiré fringe
multiplication? For a given active grating, β times as many fringes
are formed in the moiré pattern, when compared to the ordinary system in which reference and active gratings have nominally equal
frequencies. Sensitivity is increased by the factor β when a finer
reference grating is used. Naturally, β is called the *fringe
multiplication factor.*

An example is shown in Fig. 25c. The active grating applied to
the specimen had a frequency of 20 ℓ/mm before deformation; for the
reference grating, f = 400 ℓ/mm. The fringe multiplication factor
was β = 20 and 20 times as many fringes were formed, in comparison
to the ordinary moiré pattern of Fig. 25b. The increase of sensitivity is dramatically evident.

Moiré fringe multiplication is possible with β equal to odd
integers, too; see Ref.1.

Moiré Interferometry with Virtual Gratings

In Figs. 26 and 27 it is evident that two symmetrical, coherent beams of light emerge from the reference grating and

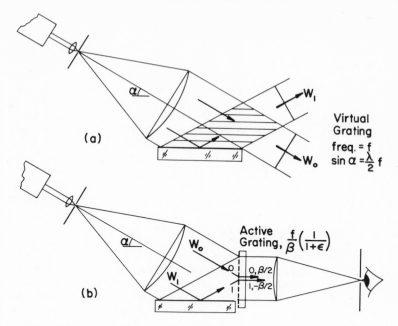

Fig. 29   (a) Formation of a virtual grating by a
mirror in a parallel beam. (b) A virtual
grating and an active grating in series
produces the same interference (or moiré)
pattern as two real gratings in series,
e.g. Fig. 27.

strike the active grating. Remember from Fig. 19 that these beams
form a virtual grating in space; it is the virtual grating
that is intercepted by the active grating. Is the reference
grating required for moiré interferometry? No, any alternative
scheme that generates a virtual grating would serve equally well.

Figure 29a shows one convenient method of generating a virtual
grating. Half of a parallel beam of coherent light is reflected by
a mirror to intersect the other half at an angle 2α. Frequency f of
the virtual grating thus formed is given by Eq. 13. The moiré interferometry setup is sketched in (b), where the virtual grating is
intercepted by a real active grating of lower frequency, nominally
f/β. If the two incident beams are named orders 0 and 1 as shown,
then the beams to be collected by the observer are order sequences
$0,\beta/2$ and $1,-\beta/2$. Of course, the result is an interference or
moiré pattern identical to that obtained with a real reference

grating of frequency f.

The method of Fig. 29b was used in an interesting investigation. Frequency of the active grating was 500 ℓ/in. With α approximately 73.5°, the virtual grating frequency was 77,000 ℓ/in. Thus, the fringe multiplication factor was 154. Angle α was adjusted to produce a null field (constant intensity field) and the active grating was rotated slightly in its plane to produce a few moiré interference fringes. The fringes had the appearance of pure two-beam interference fringes of excellent contrast.

The theoretical limit of reference frequency corresponds to α = 90° and for the wavelength of the red helium-neon laser this corresponds to 80,000 ℓ/in. Accordingly, the experiment achieved both an extra-ordinarily high fringe multiplication factor and a sensitivity of 96% of the theoretical limit.

The high fringe multiplication factor was made possible by use of an extremely high quality active grating of the bar-and-space type. Normally, irregularities along the edges of the bars create optical noise that overwhelms the low intensities of light diffracted into high diffraction orders. In this case, however, light in the ±77 diffraction orders was strong compared to the background intensity (noise) and good fringe contrast was obtained.

While the mirror scheme of Fig. 29 is convenient, many other optical systems can be employed to intersect two coherent beams and form a virtual grating.

Moiré Interferometry by Reflection

Each transmission type moire system has a reflection counterpart. The various systems will not be discussed in detail, but one particularly effective arrangement will be mentioned. It is shown in Fig. 30, where a virtual grating of frequency f is intercepted by a reflection-type active grating, initially of frequency f/2. The incoming beams are called 0 and 1 order in (b) and the paired numbers represent the order sequence after diffraction at the active grating.

In studying these sequences, note that the zeroth diffraction order in reflection is always the mirror reflection of the incident ray. Note, too, that the sign convention is changed when light propagates toward the left, with clockwise rotation from the zeroth order called positive angles.

Beams 0,1 and 1,-1 propagate to the left and are collected and observed by the eye or camera. The analogy to Fig. 29 is obvious. When frequency of the active grating is exactly f/2 and the lines

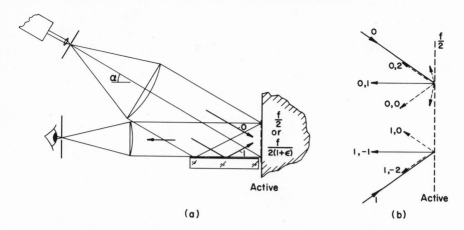

(a)   (b)

Fig. 30   (a)   Reflection counterpart of Fig. 29.
          (b)   Order sequences for reflection grating
          of frequency f/2; beams 0,1 and 1,-1 are
          collected and their interference is observed.

of the active and virtual gratings are exactly parallel, the interference pattern will be a null field. When the workpiece is subsequently deformed the fringe pattern will be a contour map of one
in-plane displacement field, given for example by

$$V = N_y/f$$

where f is frequency of the virtual reference grating.

     An example is shown in Fig. 31. The workpiece was a specimen
of a uniaxial graphite-polyamide composite with the axis of the
graphite fibers 15° from the direction of the applied tensile
forces. The loading conditions restricted the amount of rotation
and lateral translation of the ends. A cross-line phase-type reflection grating was formed on the workpiece in the upper half as
shown in (a).

     Moiré interference patterns depicting the U and V displacement
fields are shown in (b) and (c), respectively. Because of the non-
symmetric direction of the fibers, the displacements are not homogeneous (or uniform) as they would be in an isotropic tension member.
Instead they define the more complicated nonhomogeneous displacement
fields shown in the figure, which of course were the objectives
sought in these measurements.

     The active grating was a thin layer of silicone rubber, cast
between the workpiece and a cross-line furrowed mold of 600 ℓ/mm.
The mold was removed after the silicone rubber solidified.   The

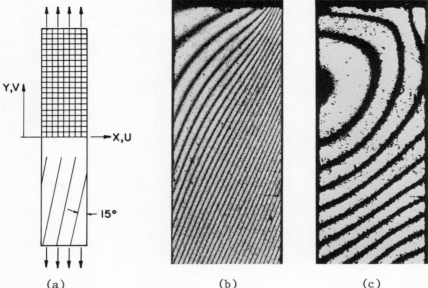

(a)   (b)   (c)

Fig. 31   Test by method of Fig. 30.   (a)   Non-isotropic tension member with cross-line grating of 600 ℓ/mm.
(b)   V-displacement field and (c) U-displacement
field.   Sensitivity is 1/1200 mm displacement per
fringe.

virtual reference grating was adjusted accurately to 1200 ℓ/mm,
as evidenced by an initial null field.

Sensitivity is 1/1200 mm or 0.833 μm per fringe.   Since fringe
order at any point can be interpolated to at least 1/5 of a fringe,
sensitivity of displacement measurements is at least 0.166 μm.

Thermal Expansion

Another recent application is measurements of thermal coefficients of expansion of composites.   The specimens were coupons taken
from thin graphite-epoxy panels, composed of a few layers laminated
together with various predetermined fiber orientations.   With such
materials, measurement of change of length is complicated by end
effects.   Since expansion is a function of measurement direction
with respect to fiber direction for a single layer, some layers
recede away from the free end while other layers bulge at the end.
Near the ends, the longitudinal strain varies from layer to layer
and the length of the specimen becomes ill-defined.

The traditional interferometric methods like that of Fig. 14
are not well suited to such non-homogeneous materials, but moiré
interferometry completely circumvents the problem.   The thin moiré

grating is applied near the center of the specimen and it does not extend to the zone affected by end conditions. Change of length with temperature manifests itself as in-plane displacements that are revealed by moiré interferometry fringe patterns.

Practical Considerations

As this technology emerged it was difficult to transfer active gratings of high frequency to the workpiece. Thus, fringe multiplication by high factors seemed to be the panacea for high sensitivity measurements. This required use of high diffraction orders from the active grating, which was uncomfortable for two reasons--light intensities emerging in high orders are tiny fractions of the incident intensity; and the noise or background intensity caused by imperfections of the active grating is usually large compared to the signal (or diffracted intensity) for moderately high-order diffractions, perhaps the 5th order and higher.

More recently, methods have emerged to replicate high frequency phase-type gratings onto the workpiece. These methods are still maturing, but it is already clear that such gratings, together with fringe multiplication by a factor of only two, is a very attractive choice. An advantage implicit in use of fringe multiplication by any even factor is normal viewing of the specimen. The choice of $\beta = 2$ prescribes that diffraction orders +1 and -1 from the active grating will be utilized. For an unblazed active grating, this means the intensity contributions will be equal, leading to two-beam interference fringes of high contrast; and it means the diffracted intensities will be greater than those for any higher fringe multiplication factor.

Replication of Gratings

The method currently in use to replicate a fine diffraction grating onto a workpiece is illustrated by the steps sketched in Fig. 32. A mold is made by first exposing a high-resolution photographic plate to two intersecting beams of coherent light. Frequency of the grating is controlled by the angle of intersection according to Eq. 13. When the plate is developed, silver grains remain in the exposed zones, while the silver is leached out in the unexposed zones. The gelatin matrix shrinks upon drying, but since it is partially restrained by the silver, shrinkage is greatest in the unexposed zones. The result is a plate with a furrowed surface which can be used as a mold.

A primer is applied to the workpiece to ensure that silicone rubber cast thereon will have good adhesion. Liquid silicone rubber is mixed with a curing agent and squeezed to a thin layer between the workpiece and mold. Upon curing, the mold releases easily

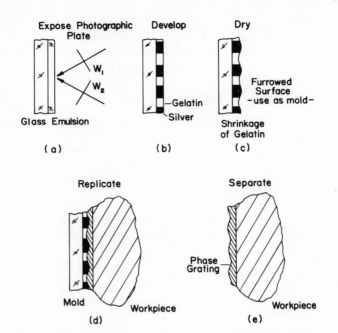

Fig. 32 Grating replication process from photographic
         exposure (a) to casting the phase grating
         onto the specimen with silicone rubber (d) &
         (e).

from the silicone rubber and the result is a replicated grating on
the workpiece.

Cross-line gratings are made in the same way, except the
photographic plate is exposed twice--once as shown and then a
repeat exposure after rotating the plate by 90°.

Overcoating with a reflective film was not done in the case of
Fig. 31. Instead, the low dielectric reflectance inherent in the
interface between the grating and air was utilized.

Further work similar to that of Walker and McKelvie[3] is in
progress in which the mold is formed by photo-resist on a glass
substrate. This will allow greater control of the depth of the
furrows. The goal is to apply a reflective coating to the mold
and transfer the furrowed shape and the reflective coating to
the workpiece--for frequencies up to 2000 ℓ/mm or 50,000 ℓ/in.

ACKNOWLEDGEMENTS

This work was supported by the National Science Foundation under Grant ENG-7824609 with Clifford J. Astill as NSF program director. This support, and the facilities and assistance of staff of the Engineering Science and Mechanics Department of VPI&SU, is greatly appreciated.

REFERENCES

1. J. Guild, The Interference System of Crossed Diffraction Gratings, Clarendon Press, Oxford, 1956.

2. D. Post, "Moiré Fringe Multiplication with a Nonsymmetrical Doubly Blazed Reference Grating," Applied Optics, Vol. 10, pp. 901-7 (1971).

3. C. A. Walker and J. McKelvie, "A Pratical Multiplied-Moiré System," Experimental Mechanics, Vol. 35, pp. 316-320 (August 1978).

BASIC WAVE ANALYSIS OF ACOUSTIC EMISSION

Robert E. Green, Jr.

Materials Science and Engineering Department
The Johns Hopkins University
Baltimore, Maryland   21218

INTRODUCTION

Acoustic emission or stress wave emission is the phenomenon of transient elastic wave generation due to a rapid release of strain energy caused by a structural alteration in a solid material. Generally these structural alterations are a result of either internally or externally applied mechanical or thermal stress. Depending on the source mechanism, acoustic emission signals may occur with frequencies ranging from several hertz up to tens of megahertz. Often the observed acoustic emission signals are classified as one of two types: burst and continuous. As shown in Fig. 1, the burst type emission resembles a damped oscillation, while the continuous type emission appears to consist of an overlapping sequence of individual bursts. The importance of acoustic emission monitoring lies in the fact that proper detection and analysis of acoustic emission signals can permit remote identification of source mechanisms and the associated structural alteration of solid materials. This information in turn can augment understanding of material behavior, can be used as a quality control method during materials processing and fabrication, and as a nondestructive evaluation technique for assessing the structural integrity of materials under service conditions.

There are a variety of sources of acoustic emission signals ranging from atomic size microstructural alterations to gross macrostructural changes such as earthquakes. A partial listing of reported sources includes movement of dislocations and grain boundaries; formation and growth of twins, crazes, microcracks, and cracks; stress-corrosion cracking; fracture of brittle inclusions and surface films; fiber breakage and delamination of composites;

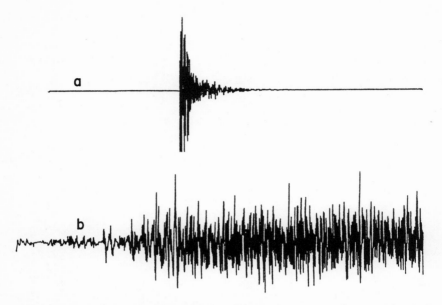

Fig. 1. Typical acoustic emission waveforms as detected with
         piezoelectric transducers, 4 msec time records:
         (a) burst type, (b) continuous type.

phase transformations; microseismic and seismic activity in geo-
logic materials.

     Although the phenomenon of acoustic emission has been the
subject of an ever increasing number of scientific investigations
and technological applications for more than 20 years[1-9], it has
not optimally fulfilled its promise as a nondestructive testing
technique since the precise characteristics of the stress waves
emitted from specific sources remain unknown. It is the purpose
of the present paper to show how proper analysis of the elastic
waves emitted from an acoustic emission source, coupled with
proper analysis of their propagational characteristics through
the workpiece can lead to positive source identification. The
validity of the theoretical calculations is proven by use of
acoustic emission sources possessing known loading features,
specimens of controlled geometry, and laser beam optical detectors.

THEORETICAL CONSIDERATIONS

     Figure 2 serves to illustrate the oversimplistic manner in
which acoustic emission is usually treated theoretically. Acoustic
emission signals propagating away from internal and surface sources

are detected by a transducer located on the surface of the test
piece. In this commonly used portrayal, the internal source emits
a single compressional spherical wave which propagates at constant
velocity in all directions. Its amplitude decreases inversely
with distance from the source only because of the expanding wavefront so that the total energy contained in the wave remains constant. The surface source also emits a single compressional
spherical wave with similar characteristics from the crack tip and,
in addition, emits a Rayleigh wave which propagates along the
surface in all directions away from the source with constant
velocity. The surface wave amplitude decreases as the square root
of the distance from the source only because of the expanding
wavefront.

Figure 3 illustrates schematically a somewhat more realistic
view of acoustic emission, although even this portrayal is still
oversimplified. The internal source emits a wave which is non-
spherical initially because of the source shape. After leaving
the source, the profile of the wavefronts continue to change because of the directional variation in wave velocities associated
with linear elastic wave propagation in anisotropic solids. Moreover, the amplitude of the wave decreases with increasing distance

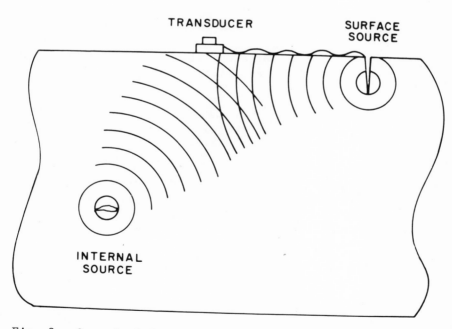

Fig. 2. Oversimplified model of acoustic emission sources.

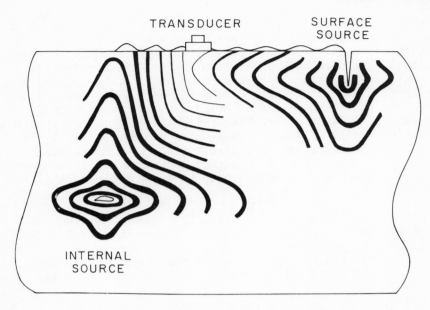

Fig. 3. More realistic simplified model of acoustic emission
         sources.

from the source, both as a result of the expanding wavefronts and
as a result of attenuation caused by a number of mechanisms which
absorb and scatter elastic waves in real materials. Among these
mechanisms are thermoelastic effects, dislocation damping, sub-
grain and grain scattering, and interaction with ferromagnetic
domain walls.

    Other factors which also influence the acoustic emission sig-
nals detected on the surface of real materials include the fact
that actual sources may emit both longitudinal and shear waves
which possess a distribution of frequencies and amplitudes. Even
in a single direction, these waves can propagate with different
wave speeds, and the higher frequency components experience stronger
attenuation than the lower frequency components. Moreover, the
elastic waves experience diffraction (beam divergence) and refraction (energy-flux deviation), mode conversion at internal and surface boundaries, and those with sufficient amplitude will be subject to non-linear effects.

EXPERIMENTAL CONSIDERATIONS

    The most simple type of acoustic emission detection system
commonly used consists of a piezoelectric transducer directly
attached to the workpiece with an acoustical impedance matching

coupling medium. The voltage output from the transducer is fed
directly into a preamplifier, which is located as close to the
transducer as possible. The preamplifier output signal is then
passed through a variable bandpass filter to a main amplifier connected to a signal analysis system, which produces an analog or
digital signal every time the amplified acoustic emission voltage
signal exceeds a selected discriminator threshold level. The most
frequently used method for evaluating structural damage by acoustic
emission monitoring is to count the signals emitted during deformation of the material and to plot the results as count rate or
total count as a function of some measure of the deformation such
as pressure, stress, strain, or number of fatigue cycles. Several
problems are associated with this simplistic approach including a
lack of knowledge of the influence of test specimen geometry on
the received signals, of the effect of the medium coupling the
transducer to the workpiece, the amplitude and frequency response
of the transducer, resonances in the active element and other
components of the transducer, and the electronic system used to
select, filter, transmit, and amplify the signals. In order to
make reliable assessment of such data, it is normally assumed that
all events producing acoustic emissions of sufficient amplitude to
be counted are equally damaging to the structure, that all damaging
events will produce acoustic emissions of sufficient amplitude to
be counted, and that each event will cause an acoustic emission
which will be counted only once and not overlap with other signals.
To make such assumptions is improper since a given high amplitude
acoustic emission signal may be produced by a single event which
causes damage to the structure or by simultaneous occurrence of
a number of small events which cause no structural damage but whose
net effect is to produce the large acoustic emission signal. A
structurally damaging event may occur, but the emissions associated
with it may be too weak, propagate away from the detecting transducer, or be of the wrong frequency to be detected. A single event
may take place with several emissions of sufficient amplitude that
the detector records several counts for one event, or a single event
may be such that the direct signal and reflected signals arrive at
the detecting transducer at different times thus resulting again
in multiple counts from a single event.

Other methods used to process acoustic emission signals are
to record the mean-square voltage, which is a measure of the energy
content, and the root-mean-square voltage, which is a measure of
the signal amplitude. In order to locate the source of an acoustic emission event, it is customary to arrange a number of transducers in a prescribed geometrical pattern on the workpiece. A
computer controlled signal analysis system is programmed to compare
arrival times of acoustic emission signals at the individual transducers and to combine this information with known wave velocities
to locate the source.

A review of the literature reveals many experiments where the detection technique did not permit recording of the unaltered acoustic emission signal characteristic of the structural defect or microstructural alteration. In other experiments proper precautions were not taken to eliminate signals caused by extraneous noise sources. In most cases where the observed acoustic emission signals were assumed to be caused by a given microstructural alteration, no metallographic or other independent evidence was presented to verify the assumption.

Figure 4(a), taken from the original work of Dunegan and Harris[10], serves as a prime example of a case where the results of a single experiment have been republished over and over again in the literature in such a fashion that the unwary reader is led to believe that the results shown represent a well-established agreement between acoustic emission measurements and theory. Figure 4(a) shows graphs of acoustic emission count rate and stress plotted as a function of strain for a 7075-T6 aluminum tensile specimen. The dashed curve superimposed on the acoustic emission data is a fit of Gilman's mobile dislocation model as a function of strain. Indeed, the agreement between theory and experiment is excellent. However, the fact that this agreement is not typical is not mentioned in the vast majority of re-publications of this figure, although this fact is clearly brought out by Dunegan and Harris in their original paper: "The results of this excellent fit of Gilman's equation to our data was very encouraging until other materials were tested and a fit attempted." Figure 4(b), also taken from their paper, shows the same set of graphs for an iron-3% silicon tensile specimen. It was impossible to obtain any fit whatsoever to Gilman's theory in this case.

The fact that Fig. 4(a) shows agreement between acoustic emission count rate and mobile dislocation density theory is probably fortuitous since not only do different sensors respond differently to the same acoustic emission signal as shown in Fig. 5, but even for a fixed detection system the recorded acoustic count rate can be made to vary over a wide range depending on choice of the detector discriminator threshold voltage setting as shown in Fig. 6. Comparison of Fig. 6 with Figs. 4(a) and (b) shows that nearly any asymmetric distribution of acoustic emission count rate can be plotted experimentally, if the correct choice of detecting transducer and discriminator threshold is made. Thus, all such plots reported to date must be regarded with some skepticism.

ACOUSTIC EMISSION SOURCE IDENTIFICATION

In order to positively identify the source of an acoustic emission signal and hence be able to make a definitive statement as to whether or not the material alteration causing the acoustic

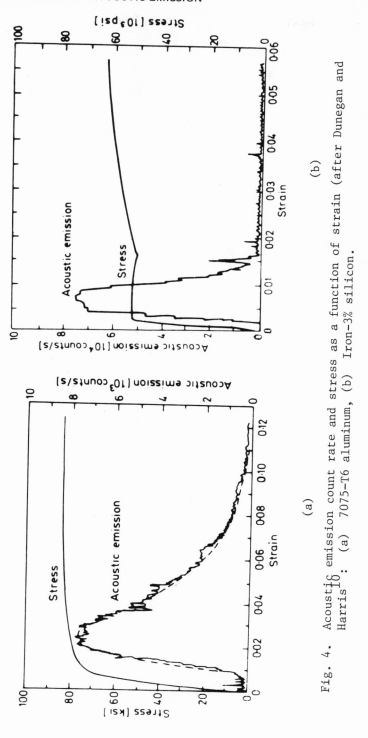

Fig. 4. Acoustic emission count rate and stress as a function of strain (after Dunegan and Harris): (a) 7075-T6 aluminum, (b) Iron-3% silicon.

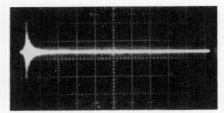

SHEAR SENSOR
1.7 MHz
EDGE MOUNTED
GAIN OF 100
HI-PASS FILTER-1.2 MHz
0.05 VOLTS/CM
0.001 SEC/CM

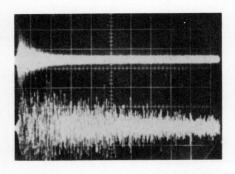

ULTRASONIC SENSOR
1 MHz CUT, SURFACE MOUNT
GAIN OF 100
HI-PASS FILTER-600 kHz
0.01 VOLTS/CM
0.001 SEC/CM

ACCELEROMETER
ENDEVCO #2213E, SURFACE MOUNT
GAIN OF 100
HI-PASS FILTER-30 kHz
0.05 VOLTS/CM
0.001 SEC/CM

Fig. 5. Different piezoelectric sensor responses to the same
acoustic emission signal (after Hutton and Ord[1]).

emission signal is harmful to the integrity of the engineering
structure, it is necessary to be able to determine the unmodified
waveform and frequency spectrum of the signal itself. Moreover,
once this determination is made, one must be able to compare the
characteristics of the acoustic emission signal in question with a
previously characterized set of acoustic emission signals recorded
from known material defects.

Prior to recent research at the National Bureau of Standards
and the Johns Hopkins University, no reliable method had been developed to solve the acoustic emission characterization problem. In
fact, numerous investigators have gone on record in the published
literature as calling the problem "unsolvable." The major reason
for such a pessimistic view is that there are several difficulties
which must be overcome in order to reliably characterize acoustic
emission signals.

First, a detector system must be used which is capable of
sensing surface displacements due to both internal and surface
acoustic emission sources which does not modify the signal because
of its own limitations. Although they have been used in the past,
and are still the detectors most often used in current practice,

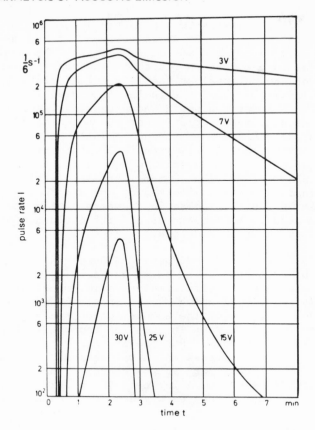

Fig. 6. Influence of detector discriminator threshold voltage
setting on shape of acoustic emission count rate versus
time curve for the same acoustic emission signal and
piezoelectric transducer (after Jax and Eisenblatter[11]).

conventional piezoelectric transducers are completely unsuited for
such measurements since they have their own amplitude, frequency,
and directional response. In addition, they "ring" at their resonance frequency and it is impossible to distinguish between the
amplitude (voltage output) excursions caused by this "ringing" and
the amplitude variations actually characteristic of the acoustic
emission source. Also, piezoelectric transducers must be coupled
to the workpiece with a material which possesses a suitable acoustical impedance match and, consequently, the waves being measured
are physically disturbed. Finally, it is uncertain just what a
piezoelectric transducer actually measures; commercial piezoelectric
transducers are known to vary markedly in their performance characteristics; and the performance characteristics which they possess
when initially purchased are known to alter detrimentally in time.

The second difficulty is to accurately determine the acoustical transfer characteristics of the workpiece. If we assume that we do have an ideal detector of surface displacements, then the question still remains as to how these surface displacements relate to the displacements of the material at the internal or distant surface source of the acoustic emission signals. The need is for the transfer function which describes how a known source signal is modified by the propagational characteristics of the material as the elastic waves leave the source and pass through a solid body of prescribed geometry. Determination of this transfer function is a necessity in order to answer the real question which is the inverse problem. Assuming that all surface displacements are accurately known, and the direct transfer function also known, one can then solve the inverse problem of source displacement determination and hence identify the source mechanism and assess its influence on the structural integrity of the workpiece.

Considerable progress in acoustic emission waveform determination was made by Breckenridge et al.[12] who developed a method for obtaining the waveforms of simulated acoustic emission sources which were unmodified by ringing of the specimen and transducer. Their technique was based on the comparison of two signals at the transducer, one from the event in question and one from an artificial event of known waveform. For the standard-source event they chose a step function of stress which was realized experimentally by fracturing a glass capillary at the center of one flat surface of large cylindrical aluminum alloy transfer block. In order to have a broadband detector they used a dc-biased capacitive transducer with an air-gap dielectric. They found that measurements of the waveform, both on the same surface as the source and on the opposite surface yielded results that agreed remarkably well with theoretical predictions. An important conclusion from their work was that no difference was observed between the shape of the vertical displacement at the epicentral point produced by a source event on the surface and that produced by a similar event in the interior of a solid.

Subsequently, Hsu et al.[13,14] numerically computed the displacement as a function of time at an arbitrary point on an infinite plate due to a point source force function. Nearly perfect experimental agreement with their theory was obtained by using a reproducible step-function stress release pulse as a simulated acoustic emission signal and a wide band capacitive transducer as a sensor.

OPTICAL DETECTION OF ACOUSTIC EMISSION SIGNALS

In order to optimally solve the detector problem several laser beam optical systems for detection of acoustic emission signals have been developed[15-32]. These optical detectors offer many

important advantages over piezoelectric transducers for monitoring
acoustic emission events. Among these advantages are: direct contact with the test specimen is unnecessary; no acoustical impedance
matching couplant is required; the waveform and frequency spectrum
of the acoustic emission event is not modified by the optical probe;
optical probes have inherent broad flat frequency responses; they
can probe internally in transparent media; they can be used to make
measurements on extremely hot and extremely cold materials and in
other environments hostile to piezoelectric transducers; since the
focused laser beam diameters are typically only a few hundredths of
a millimeter, they can probe very close to a slip band, mechanical
twin, included particle, crack, or other material defect.

An optical acoustic emission detector can be used as a point
probe to measure the waveforms directly at the site of an internal
source in a transparent solid. It can be used to measure the waveforms as a function of position in the solid as well as on its surface. Thus correlation can be achieved between the acoustical
characteristics of the source and the detected surface signals.
A laser beam acoustic emission detector system completely solves
the aforementioned detector difficulty and permits progress to be
made on the transfer function problem.

Optical probes have already been used to measure mechanical
twinning of cadmium, indium, tin, and zinc; cracking of glass by
both thermal and mechanical means; stress-corrosion cracking in
E4340 steel and 7039 aluminum; allotropic transformation in iron;
and plastic deformation and fracture of 304L stainless steel.

ACOUSTIC EMISSION SIGNAL CAPTURE AND PROCESSING

Conventional procedures for capturing acoustic emission waveforms and subsequent processing, such as Fourier analysis in order
to determine the frequency spectrum, usually involve use of a
"broadband" piezoelectric transducer, a video tape recorder, and a
commercially available spectrum analyzer[33]. The gross deficiencies
of piezoelectric transducers for this purpose have already been
enumerated and will be illustrated by a specific example in the
next section. Although a video tape recorder has proven to be an
economical method of almost continuous recording of acoustic
emission signals over time periods as long as several hours, it is
not an optimum device for faithful recording and reproduction of
acoustic emission waveforms. It has limited bandwidth (4 MHz),
clips negative signals more than positive ones, and requires use
of a gate in order to remove signal artifacts due to synchroniza-
tion pulses internally generated by the recorder and alternation
of the video recording heads.

The best system currently available for reliable recording
and reproduction of individual acoustic emission waveforms is a

transient recorder which has a high digitizing speed and a long
time window coupled to a digital magnetic disc or tape recorder.
The magnetically stored signal serves as input to a high speed
digital computer for signal processing including fast Fourier
transforms for spectral analysis.

VERIFICATION OF SOURCE IDENTIFICATION SYSTEM CAPABILITY

In order to assure that a selected experimental acoustic
emission source identification system can be successfully used to
detect and record the unknown waveforms of acoustic emission
signals characteristic of real structural or microstructural
materials alterations, it must first be shown that the system is
capable of faithful detection and recording of waveforms produced
under carefully controlled experimental conditions which correspond to known theoretical waveform solutions. Although this has
been successfully accomplished using capacitive transducers[12-14]
and optical detectors[26-29], there has not been a single case of
such an accomplishment using piezoelectric transducers.

Figure 7 shows the theoretically computed vertical surface
displacement versus time record for a step-function time dependence
point source load applied to the surface of a homogeneous isotropic
linear elastic half-space. As mentioned previously, Breckenridge
et al.[12] used a capacitive transducer to measure the vertical surface displacement due to a rapid unloading pulse caused by fracture
of a glass capillary tube against a specially designed test block.
The results of this measurement, as can be seen reproduced in
Fig. 12 of the paper by Weisinger[35], show most remarkable agreement
with the theoretical predictions. When the surface displacement
was measured over a longer time period of several milliseconds,
the waveform shown in Fig. 8 was obtained. The actual waveform
characteristic of the source can still be seen at the far left of
the figure, while the subsequent large oscillations are deemed to
be due to reverberations in the test block and support structure.
Similar results were obtained by Hsu and Hardy[14] for epicenter
displacements produced by breaking glass capillaries on an aluminum
plate.

Figure 9 compares 200 sec waveforms from fracture of a glass
capillary against an aluminum test block as detected (a) optically
and (b) piezoelectrically. The portion of the optically detected
signal to the left of the vertical dashed line agrees within experimental accuracy with theoretical predictions of the displacement
associated with the source. An arrow indicates an enlarged view
of this portion of the waveform. Even longer time records, 2 msec,
of the same two waveforms are presented in Fig. 10. As can be seen
in Figs. 9 and 10, the piezoelectric transducer introduces a multi-
tude of oscillations into the detected waveform and does not faithfully reproduce the waveform due to surface displacements charac-

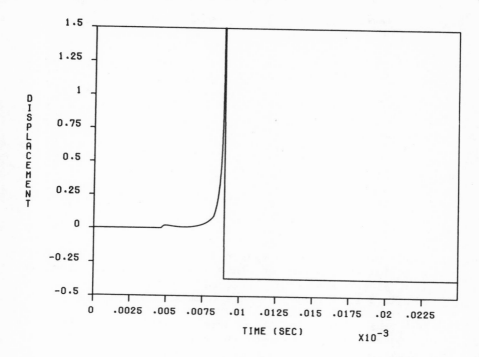

Fig. 7. Theoretical vertical surface displacement versus time
         record for a step-function time dependence point source
         load applied to the surface of a homogeneous isotropic
         linear elastic half-space.

teristic of the acoustic emission source.

WAVEFORM AND FREQUENCY MODIFICATION BY PROPAGATION

     As has been shown in the previous section, even when a capaci-
tive transducer or optical probe is used to faithfully detect the
surface displacement waveform characteristic of an acoustic emission
source, the waveform, and hence the frequency spectrum, appears to
undergo severe modification as it reflects back-and-forth in an
actual specimen. In order to absolutely prove or disprove this
hypothesis, it is necessary to compute theoretically the waveforms
and frequency spectra to be expected from a variety of specific
combinations of well-characterized sources and specimens. It is
desired to calculate the particle motion both inside and on the
surface of a specimen of known material and geometry as a result
of a known applied force. A schematic illustration of this problem

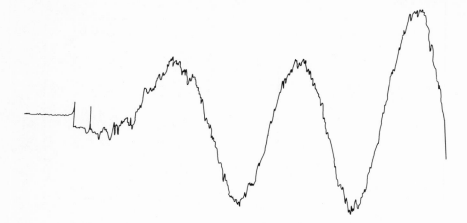

Fig. 8.   Capacitive transducer measured vertical surface displace-
          ment due to fracture of glass capillary against aluminum
          test block, several millisecond time record
          (F. R. Breckenridge, private communication).

in which only the geometry of the specimen is changed is shown in
Fig. 11.   A survey of the state-of-the art of progress in this
important area is given by Weisinger[35].

RECENT EXPERIMENTAL MEASUREMENTS WITH MICRO-TENSILE SPECIMENS

       Although the literature is saturated with claims of acoustic
emission source identification due to a variety of microstructural
alterations in real materials, all of these measurements have been
made on large specimens with piezoelectric transducers either located on the specimens at locations remote from the source or,
even worse, located on specimen grips or other such supports.   It
should be obvious from the considerations of the previous sections
why the present author views such measurements with skepticism.

       In an effort to measure the acoustic emission signals as
close to an actual microstructural alteration source as possible
in a real material experiments were initiated with micro-tensile
test specimens possessing very small gauge sections as shown in
Fig. 12.   This reduction in the volume of material from which
acoustic emissions could be produced during tensile deformation
offers the following advantages: permits placement of the optical
probe extremely close to the source; reduces the number of potential sites for acoustic emission generation; reduces the overlap
between signals since fewer will be generated; reduces the number
of potential loss or attenuation sites present.   Using 304L stainless steel specimens of this geometry, coupled with a laser beam
optical probe, and an ultra-quiet pneumatic tensile load system,

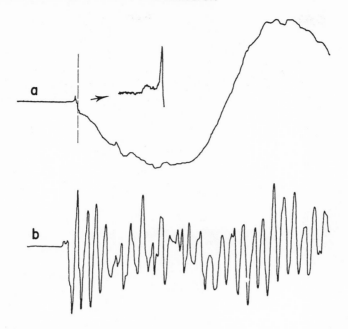

Fig. 9. Comparison of waveforms from fracture of a glass capillary
         as detected (a) optically and (b) piezoelectrically,
         200 sec time records. The portion of the optically
         detected signal to the left of the vertical dashed line
         agrees with theoretical predictions of the displacement
         associated with the source. Arrow indicates enlarged view
         of this portion of waveform. (after Djordjevic and
         Green[26]).

Bruchey[32] found that the specimens failed by void nucleation, growth,
and coalescence. Examination of the fracture surface of each failed
specimen in a Scanning Electron Microscope (SEM) revealed at a magnification of 500X or greater that each void contained a rather large
intermetallic particle. At the base of some of these voids were
particles that had cracked in a brittle manner (Fig. 13). By scanning the entire fracture surface of the steel gauge section and
counting the number of broken intermetallic particles, Bruchey
found a near one-to-one correspondence between fractured intermetallic particle count and the number of acoustic emission bursts
in the 7.5-8.5 MHz regime. Although fracture of intermetallics
have been offered by a number of authors in a number of materials

INTERFEROMETER

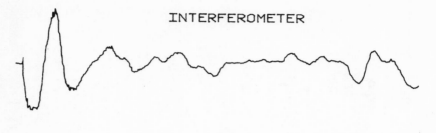

A.E. TRANSDUCER

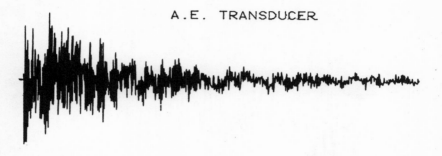

Fig. 10.   Longer time records, 2 msec, of same two waveforms as
           shown in Fig. 9 (after Djordjevic[28]).

as the source of emissions, this is believed to be the first time
that direct counts, rather than statistical counts, support this
hypothesis. It should be noted that the small gauge size and flat
frequency response of the optical probe up to at least 10 MHz were
both necessary for successful completion of such measurements.

CONCLUSIONS

     Proper analysis of the elastic waves emitted from an acoustic
emission source coupled with proper analysis of their propagational
characteristics through the workpiece can lead to specific source
characterization. Both capacitive transducers and optical laser
beam detectors have proven capable of detecting surface displacements due to acoustic emission sources in a reliable fashion;
piezoelectric transducers are completely unsatisfactory. Additional
theoretical calculations and reliable experimental measurements need
to be made on a variety of specific combinations of well-characterized sources and specimens. It is desirable to conduct additional experiments where optical probes are used to detect acoustic
emission signals from micro-tensile specimens and sheet specimens
made from a variety of materials and to use optical and electron
microscopy to positively correlate observed sources with measured
acoustic emission signals.

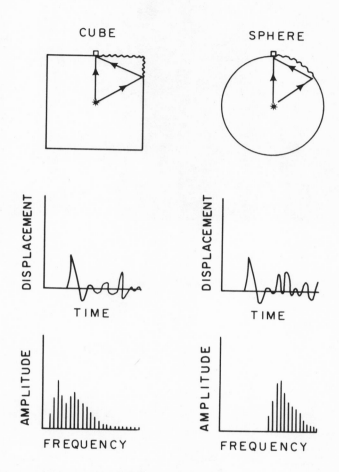

Fig. 11. Schematic illustration of waveform and frequency spectrum
modification of acoustic emission signal due to specimen
geometry (R. Weisinger, private communication).

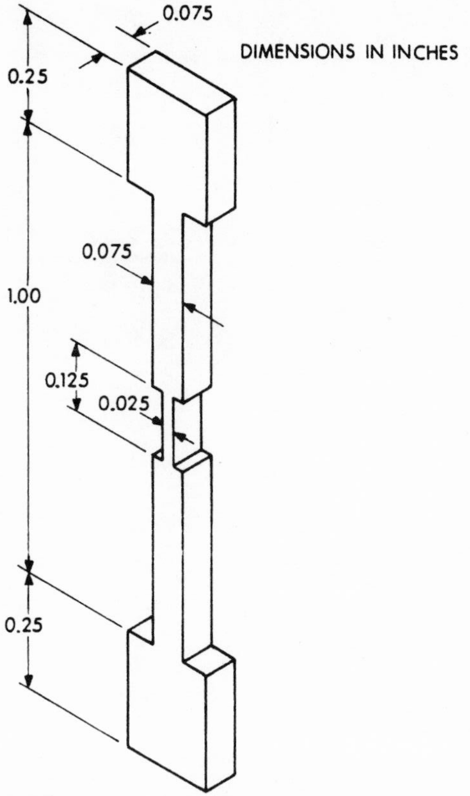

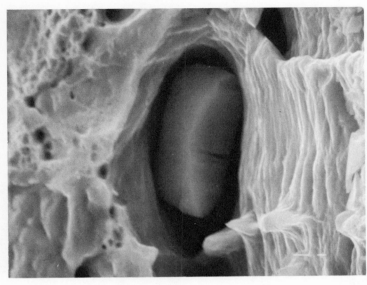

Fig. 13. SEM photograph of 304L stainless
steel micro-tensile specimen
fracture surface showing void
containing cracked intermetallic
particle (4500 X) (after Bruchey32).

DIMENSIONS IN INCHES

0.075

0.25

1.00

0.075

0.125

0.025

0.25

Fig. 12. Dimensional drawing of micro-tensile
specimen (after Bruchey32).

ACKNOWLEDGEMENTS

    The initial phases of this work were supported in part by the
U. S. Army Research Office, while the current effort is partially
supported by the U. S. Naval Sea Systems Command. A special note
of appreciation in this regard is due Dr. George Mayer and Dr.
H. H. Vanderveldt. Thanks are due Professor C. H. Palmer, Dr.
Steven E. Fick, Dr. B. Boro Djordjevic, Dr. William J. Bruchey, Jr.,
and Mr. Richard Weisinger for their technical assistance. Grateful
appreciation is due Mrs. Alicia Falcone for her care and effort in
typing the manuscript and to Mr. Richard Weisinger for his assistance in preparing the camera-ready copy.

REFERENCES

1. Hutton, P. H. and Ord, R. N., "Acoustic Emission," in Research
    Techniques in Nondestructive Testing, Vol. 1, pp. 1-30,
    R. S. Sharpe, ed., Academic Press, London and New York (1970).

2. Acoustic Emission, R. G. Liptai, D. O. Harris, and C. A. Tatro,
    eds., ASTM STP 505, American Society for Testing and Materials,
    Philadelphia, Pa. (1972).

3. Pollock, A. A., "Acoustic emission, A review of recent progress
    and technical aspects," in Acoustics and Vibration Progress,
    Vol. 1, pp. 53-84, R. W. B. Stephens and H. G. Leventhall, eds.,
    Chapman and Hall, London (1974).

4. Spanner, J. C., Acoustic Emission, Techniques and Applications,
    Intex Publishing Co., Evanston, Illinois (1974).

5. Lord, A. E., Jr., "Acoustic Emission," in Physical Acoustics,
    Vol. 11, pp. 289-353, W. P. Mason and R. N. Thurston, eds.,
    Academic Press, New York (1975).

6. Monitoring Structural Integrity by Acoustic Emission,
    J. C. Spanner and J. W. McElroy, eds., ASTM STP 571, American
    Society for Testing and Materials, Philadelphia, Pa. (1975).

7. Acoustic Emission, R. W. Nichols, ed., Applied Science Pub-
    lishers Ltd., London (1976).

8. Proceedings First Conference on Acoustic Emission/Microseismic
    Activity in Geologic Structures and Materials, H. R. Hardy, Jr.,
    and F. W. Leighton, eds., Trans Tech Publications, Clausthal,
    Germany (1977).

9. Droillard, T. F., Acoustic Emission, a Bibliography with Ab-
    stracts, F. J. Laner, ed., IFI/Plenum, New York (1979).

10. Dunegan, H. and Harris, D., "Acoustic Emission - A New Non-destructive Testing Tool," Ultrasonics 7, 160-166 (1969).

11. Jax, P. and Eisenblatter, J., "Acoustic Emission Measurements During Plastic Deformation of Metals," Information 15 Battelle Frankfurt, pp. 2-8 (April 1973).

12. Breckenridge, F. R., Tschiegg, C. E., and Greenspan, M., "Acoustic emission: some applications of Lamb's problem," J. Acoust. Soc. Amer. 57, 625-631 (1975).

13. Hsu, N. N., Simmons, J. A., and Hardy, S. C., "An Approach to Acoustic Emission Signal Analysis - Theory and Experiment," Materials Evaluation 35, 100-106 (1977).

14. Hsu, N. N. and Hardy, S. C., "Experiments in Acoustic Emission Waveform Analysis for Characterization of AE Sources, Sensors and Structures," in Elastic Waves and Non-Destructive Testing of Materials, Y. H. Pao, ed., AMD - Vol. 29, pp. 85-106, The American Society of Mechanical Engineers, New York (1978).

15. Palmer, C. H. and Green, R. E., Jr., "Optical Probing of Acoustic Emission Waves," in Nondestructive Evaluation of Materials, J. J. Burke and V. Weiss, eds., Sagamore Army Materials Research Conference, Raquette Lake, New York (1976), Conference Proceedings 23, 347-378, Plenum Press, New York (1979).

16. Green, R. E., Jr. and Palmer, C. H., "Optical Detection of Acoustic Emission Waves," Proceedings of the Ninth International Congress on Acoustics," Madrid, Spain (July 1977).

17. Palmer, C. H. and Green, R. E., Jr., "Optical Detection of Acoustic Emission Waves," Applied Optics 16, 2333-2334 (1977).

18. Palmer, C. H. and Green, R. E., Jr., "Materials Evaluation by Optical Detection of Acoustic Emission Signals," Materials Evaluation - Research Supplement 35, 107-112 (1977).

19. Palmer, C. H. and Green, R. E., Jr., "Optical Detection of Acoustic Emission Signals," Proceedings of ARPA/AFML Review of Progress in Quantitative NDE Meeting, Cornell University, Ithaca, New York (June 1977), AFML-TR-78-55 (May 1978).

20. Palmer, C. H., "Optical Detection of Ultrasonic Transients," Proceedings of 1978 Conference on Information Sciences and Systems, The Johns Hopkins University, Baltimore, Maryland, pp. 559-563 (1978).

21. Kline, R. A., "Optical and Piezoelectric Detection of Acoustic Emission Signals," Ph.D. Thesis, Mechanics and Materials Science Department, The Johns Hopkins University, Baltimore, Maryland (1978).

22. Kline, R. A., Green, R. E., Jr., and Palmer, C. H., "A Comparison of Optically and Piezoelectrically Sensed Acoustic Emission Signals," J. Acoust. Soc. Am. 64, 1633-1639 (1978).

23. Palmer, C. H., "Optical Measurement of Acoustic Emission at High and Low Temperatures," Proceedings of ARPA/AFML Review of Progress in Quantitative NDE Meeting, La Jolla, California (July 1978), AFML-TR-78-205 (January 1979).

24. Palmer, C. H. and Fick, S. E., "New Optical Instrument for Acoustic Emission," IEEE Southeastcon, Roanoke, Virginia, pp. 191-192 (April 1979).

25. Green, R. E., Jr., Djordjevic, B. B., Palmer, C. H., and Fick, S. E., "Laser Beam Detection of Ultrasonic and Acoustic Emission Signals for Nondestructive Testing of Materials," Proceedings of ASM Conference on Applications of Lasers to Materials Processing, pp. 161-175, Washington, D. C. (April 1979), American Society for Metals, Metals Park, Ohio (1979).

26. Djordjevic, B. B. and Green, R. E., Jr., "High Speed Digital Capture of Acoustic Emission and Ultrasonic Transients as Detected with Optical Laser Beam Probes," Proceedings of Ultrasonics International 79 Conference, Graz, Austria (May 1979), pp. 82-87, IPC Science and Technology Press Ltd., Guildford, England (1979).

27. Palmer, C. H. and Green, R. E., Jr., "Optical Probing of Acoustic Emission Waves," Final Report, U. S. Army Research Office, The Johns Hopkins University (August 1979).

28. Djordjevic, B. B., "Digital Waveform Recording and Computer Analysis of Acoustic Emission and Ultrasonic Transients Detected by Optical and Piezoelectric Probes," Ph.D. Thesis, Mechanics and Materials Science Department, The Johns Hopkins University, Baltimore, Maryland (1979).

29. Green, R. E., Jr., "Acoustic Emission," Encyclopedia of Science and Technology, McGraw-Hill, New York (to be published 1980).

30. Kline, R. A., Green, R. E., Jr. and Palmer, C. H., "Acoustic Emission Waveforms from Stress-Corrosion Cracking of Steel: Experiment and Theory (submitted for publication in J. Appl. Phys.).

31. Peterlin, A., Djordjevic, B. B., Murphy, J. C., and Green,
     R. E., Jr., "Acoustic Emission During Craze Formation in
     Polymers (submitted for publication in J. Appl. Phys.).

32. Bruchey, W. J., Jr., "Optical Probing of Acoustic Emission
     During Deformation of Micro-Tensile Specimens," Ph.D. Thesis,
     Materials Science and Engineering Department, The Johns
     Hopkins University, Baltimore, Maryland (1980).

33. Graham, L. J. and Alers, G. A., "Spectrum Analysis of Acoustic
     Emission in A-533-B Steel," Materials Evaluation 32, 31-37
     (1974).

34. Weisinger, R., "Acoustic Emission and Lamb's Problem on a Homo-
     geneous Sphere," Masters Essay, Mechanics and Materials Science
     Department, The Johns Hopkins University, Baltimore, Maryland
     (1979).

35. Weisinger, R., "Determination of Fundamental Acoustic Emission
     Signal Characteristics," Proceedings of Conference on Mechanics
     of Nondestructive Testing, Virginia Polytechnic Institute and
     State University (September 1980).

A SURVEY OF ELECTROMAGNETIC METHODS OF NONDESTRUCTIVE TESTING*

W. Lord

Electrical Engineering Department
Colorado State University
Fort Collins, Colorado 80523

SUMMARY

Electromagnetic methods of nondestructive testing (NDT) can be classified according to their mode of specimen excitation:

· Direct current excitation results in <u>active</u> leakage fields around defects in ferromagnetic materials.

· <u>Residual</u> leakage fields around defects in ferromagnetic materials occur after removal of the dc excitation current.

· Alternating current excitation of conducting materials results in induced <u>eddy currents</u> which are sensitive to a wide variety of specimen properties.

All of these techniques, when used in conjunction with an appropriate test probe, can be applied to the problems of testing metals nondestructively. Each technique has its own particular field of application and many are used on a daily basis in the energy, transportation and aerospace industries. The classification covers not only a broad frequency spectrum (from 0Hz into the MHz range) but also a large number of individual testing techniques associated with each sub group. As many of the electromagnetic NDT phenomena have been known and studied since the end of the last century, it should come as no surprise to the reader to find that the literature associated with this area is both rich and diverse with regard to applications and theoretical

---
*This work has been supported in part by the Colorado Energy Research Institute, the Army Research Office and the Electric Power Research Institute.

modeling.

Despite the longevity of the subject matter, industry's demands for better performance and reliability from metal components have caused increased interest in improving the state of the NDT art. This has resulted in a growing number of studies into the modeling of basic electromagnetic field/defect interactions and into techniques for improving the evaluation of test signals.

The major objective of this paper is to provide a survey of the topic by describing both practical and theoretical developments to-date and indicating current and future trends, thus characterizing the general philosophy of the field. An extensive bibliography is included with the paper which should enable the reader to obtain further in-depth information concerning most aspects of electromagnetic NDT techniques, and which also serves to indicate the general resurgence of interest in the field.

INTRODUCTION

Nondestructive testing (NDT) techniques rely for their operation on the interaction of an energy source with the material parameter of interest and the subsequent measurement of the perturbed field by an appropriate transducer. Electromagnetic NDT techniques are no exception. Indeed, as electromagnetism refers to the production of a magnetic field by an electric current, and as magnetic field strength is proportional to the reluctance "seen" by the source of magnetization, and as electric current flow depends upon material conductivity, these NDT techniques find a whole host of applications from sizing metal products, detecting surface and subsurface flaws, measuring temperature and flux density to sorting various alloys. Such flexibility is not without its disadvantages, however, as Hughes[1] reported in his classic paper on eddy current studies in 1897 wherein he noted the extreme sensitivity of his apparatus to external disturbances. The wide range of applications and operating modes and the broad frequency spectrum associated with electromagentic NDT methods make the task of surveying the subject very difficult; additional frustration surfaces when one considers the subject matters longevity and the proclivity of testing personnel to classify all electromagnetic NDT phenomena separately into magnetic particle, flux leakage and eddy current categories, treating each category as an entity unto itself. This paper classifies electromagnetic NDT methods by excitation mode and shows, for ferromagnetic materials, how the excitation modes and hence the various electromagnetic NDT methods themselves are related via the materials B/H (magnetization) characteristic; in so doing a survey of theoretical modeling and applications is given. Typical examples are taken from the author's own NDT experiences.

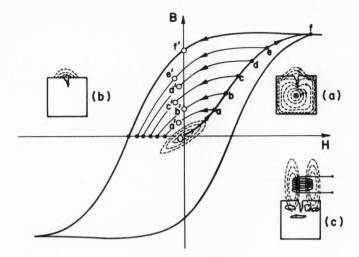

Fig. 1.   Interrelationship of a) active, b) residual and c) eddy current NDT methods via the specimens hysteresis (B/H) characteristic.

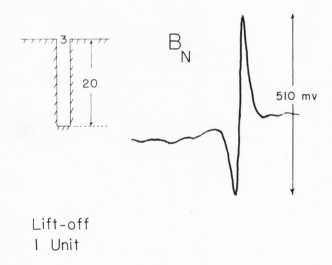

Fig. 2.   Typical active "leakage field profile" around a rectangular slot in a steel bar magnetized axially with a steady direct current.

ACTIVE DC EXCITATION

    At the lower end of the frequency spectrum, dc current can
be used as the excitation source to set up an internal magnetic
field within the specimen. The strength of this dc (magnetostatic)
field can be found from the Maxwell-Ampere law

$$\oint_c \bar{H} \cdot d\bar{\ell} = \iint_s \bar{J} \cdot d\bar{s} \qquad\qquad (1)$$

which states that the magnetizing force $\bar{H}$ acting around a closed
contour c is equal to the current enclosed by the contour. Knowing H from this relationship, B can be found directly as

$$B = \mu_o \mu_R \, H \qquad\qquad (2)$$

assuming a homogeneous, isotropic medium. The situation is depicted graphically in Fig. 1 from which it can be seen that each
element of a ferromagnetic material excited by an active dc current
works at a unique point on the nonlinear magnetization curve in the
first quadrant of the materials B/H characteristic.

    From a practical standpoint, it is the materials domain
behavior, summarized by equation 2 and illustrated in Fig. 1,
which leads to the following usable NDT phenomnea:

   · Flaw detection - surface and subsurface flaws represent
     an increase in reluctance to the magnetic flux lines
     and some of them "leak out" as shown in Figs. 1 and 2.
     Leakage flux can be detected by Hall probes, magnetic
     particles, moving coils and even recorded on magnetic
     tape[2,3]. A typical Hall probe "leakage field profile"
     is shown in Fig. 2. Such techniques are used widely
     in oil and gas industries for testing pipelines for
     corrosion and cracks, in the welding of steel products
     for detecting flaws in machined steel components and
     in the detection of cracks in railroad wagon wheels
     to name but a few applications[4-13].

   · Sizing and Clearance Measurements - the reluctance of
     a magnetic circuit is given by

$$R = \ell / \mu_o \mu_R \, A \qquad\qquad (3)$$

     where ℓ is the effective length of the flux paths and A
     corresponding cross-sectional area. Because of the
     relative permeability term μ_R, flux paths through air
     have a much higher reluctance than those through ferro-
     magnetic materials, consequently length or area varia-
     tions in an air gap can be monitored directly by measuring

the strength of the magnetic field with an appropriate
flux sensitive transducer. A typical example of such
a simple reluctance probe is given in Fig. 3 which shows
details of a test rig for measuring the crevice gap
clearance (distance between the tube and support plate)
in a steam generator unit. A typical probe output
signal is shown in Fig. 4. Repeated tests with different support plates have shown that the magnitude of the
probe signal is related directly to the crevice gap
clearance[14].

Metallurgical properties, stress and temperature measurement, metals sorting - the magnetization curve in Fig. 1
has a nonlinear shape due to the behavior of domains in
the ferromagnetic material. Such behavior is affected by
grain structure, hardness, stress, temperature and other
variables. A variety of tests have been devised to take
advantage of these magnetic phenomena and provide non-
interfering measurements of the variables[15-19].

In flaw detection work, the active dc fields are set up by
passing a current through the specimen. Care has to be taken to
ensure that the direction of the resultant field is normal to the
defect orientation. Techniques have been developed for inducing
a magnetic field within a specimen by non-contacting means[20-22].
This eliminates the problems associated with handling the high dc
currents needed to magnetize large ferromagnetic specimens.

RESIDUAL MAGNETIZATION

When an active dc excitation current is removed from a ferromagnetic specimen the working point of each internal element re-
laxes from its position on the initial magnetization curve shown
in Fig. 1, to either a remanent flux density point on the H=0 axis
or, in the case of those elements close to an open defect, to a
point in the second (demagnetization) quadrant of the materials
B/H loop. This effect is shown in Fig. 1 and the net result is
that a surface defect behaves somewhat like a permanent magnet
with residual leakage flux from the defect surfaces extending into
the surrounding atmosphere. Such defect leakage fields can be
detected using any of the flux sensitive devices mentioned in the
active excitation section; traditionally, however, magnetic particles have been used for this purpose[3,4,23,24].

Fig. 5 shows details of a test rig used to study the residual
leakage field profiles around rectangular slots using a Hall probe
and Fig. 6 gives a typical output signal. Apart from the obvious
differences in signal magnitude between residual and active leakage
field profiles, there are also differences in signal shape[25].

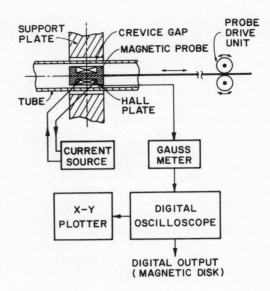

Fig. 3.   Steam generator test rig for measurement of crevice
          gap clearance using a dc variable reluctance probe.

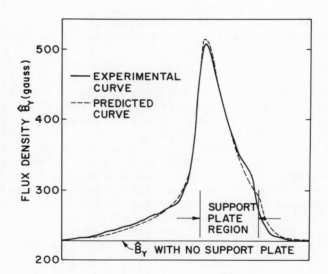

Fig. 4.   Hall element output signal for the probe passing through
          a carbon steel support plate. Signal peak occurs when
          the probe is perfectly aligned with the support plate and
          "sees" the least reluctance. Dashed curve corresponds to
          the finite element prediction of probe output.

Fig. 5. Experimental test rig used for studies of residual leakage fields around rectangular slots in steel specimens.

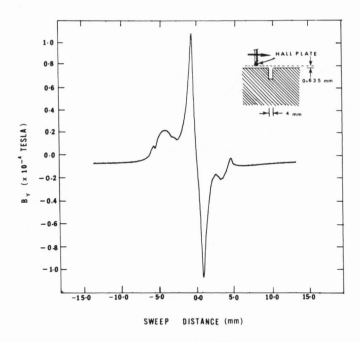

SWEEP   DISTANCE (mm)

Fig. 6. Typical residual leakage field profile around a rectangular slot in a steel bar <u>after</u> removal of dc excitation
current.

Major advantages of the magnetic particle technique for detecting defects relate to the method's inherent simplicity and sensitivity. Unfortunately, the results tend to be qualitative rather
than quantitative and do not lend themselves readily to automation
in the signal processing sense. The technique is used extensively
for detecting surface defects in both steel stock and machined
parts as well as for weld testing[3,4,26,27].

ALTERNATING CURRENT EXCITATION

Active and residual magnetization techniques can only be used
with ferromagnetic materials because of the relative permeability
term in equation 2. Most nonmagnetic materials have a relative
permeability close to that of air and hence magnetostatic leakage
fields around defects in such materials do not exist. Consequently,
except at very low frequencies the mode of operation for eddy current testing techniques is somewhat different than that for active
and residual methods. When a coil carrying an alternating current
is placed above a conducting specimen, eddy currents are induced
in the material by the sinusoidally pulsating field set up by the
excitation coil. The magnitude of these eddy currents is governed
by the equations

$$\oint_c \bar{E} \cdot d\bar{\ell} = - \iint_s \dot{\bar{B}} \cdot d\bar{s} \tag{4}$$

$$\bar{J} = \sigma \bar{E} \tag{5}$$

and the eddy currents themselves set up a magnetic flux which
interacts with the main flux.

Specimen defects, changes in conductivity, dimension variations, etc. all affect the flow of eddy currents and this can be
measured as a change in impedance of the excitation coil or a
nearby search coil. The depth of penetration δ, of the eddy currents can be varied by changing the excitation frequency f, a
degree of freedom unobtainable with active or residual dc fields.

$$\delta = \frac{1}{\sqrt{\pi f \mu \sigma}} \tag{6}$$

where $\mu = \mu_o \mu_R$ and σ is the conductivity of the specimen.

Eddy current techniques are used widely throughout industry
for detecting defects in metal products; sizing tube, rod, plate
and coating dimensions; and measuring electrical conductivity, to
name just the more common applications[28-34]. Such methods are
applicable to all conducting materials. In the case of ferromagnetic specimens the working point of each element of the material
will trace a complete hysteresis loop in the B/H plane (see Fig. 1)

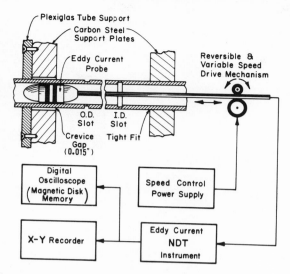

Fig. 7. Eddy current test rig using a differential probe for studying impedance plane trajectories associated with defects in steam generator tubing.

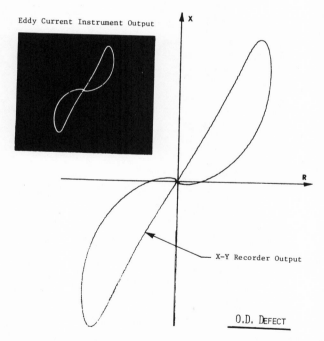

Fig. 8. Typical impedance plane trajectories from the test rig of Fig. 7. a) O.D. tube slot signal.

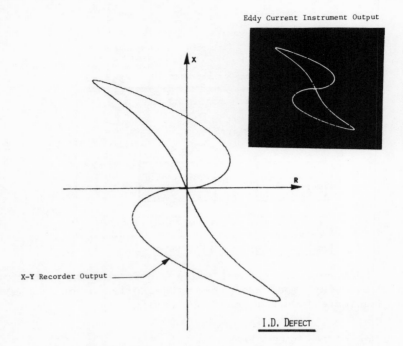

Fig. 8b).   I.D. tube slot signal

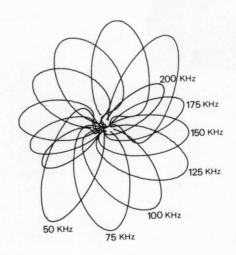

Fig. 8c).   Support plate signal.

whose magnitude and shape will depend upon the strength of the induced eddy currents at that point.

Comments with regard to eddy current methods are limited in this paper to the intermediate frequency range (up to the low MHz region) where diffusion equations can be used to describe the behavior of sinusoidal excitation and induced currents and fields. Readers should be aware, however, that excitation frequencies beyond this range are being considered for measuring surface defects in metals by microwave scattering[35,36], and also that pulsed electromagnetic fields have been used with success for a variety of NDT applications[37].

Figure 7 shows a typical eddy current testing arrangement for studying the impedance plane trajectories (Lissajous patterns) associated with defects in steam generator tubing and Fig. 8 gives typical differential probe output signals.

MODELING FIELD/DEFECT INTERACTIONS

Demands for better performance and reliability from metal components and structures have caused increased interest to be generated in improving the state of the NDT art. This has resulted in a growing number of studies into the modeling of basic electromagnetic field/defect interactions, both to improve understanding of electromagnetic NDT phenomena, and to provide training data for automated signal processing equipment.

The problems of predicting active and residual leakage fields around defects, and impedance plane trajectories for eddy current probes are complicated by the fact that the equations describing the fields in the vicinity of a defect are nonlinear, three dimensional, partial differential equations with awkward boundary conditions. Analytically, solutions are all but impossible to obtain for anything but the simplest of geometries, and even then, only by making such simplifying assumptions that the techniques do not lend themselves well to the defect characterization problem.

For active and residual leakage fields one analytical approach has been to model rectangular surface slots by strip dipoles[38-39]. This leads to equations for the x and y components of the magnetizing force above the slot of the form:

$$H_x = 2\sigma_s [arctg\frac{h(x+b)}{(x+b)^2+y(y+h)} - arctg\frac{h(x-b)}{(x-b)^2+y(y+h)}] \tag{7}$$

$$H_y = \sigma_s \ln\frac{[(x+b)^2+(y+h)^2][(x-b)^2+y^2]}{[(x+b)^2+y^2][(x-b)^2+(y+h)^2]} \tag{8}$$

where b is one half the slot width, h is the slot depth and σ_s is
the fictitious "magnetic charge density" over the surface of the
slot. Figs. 9 and 10 show typical H_x and H_y components as predicted by equations 7 and 8 for the slots whose dimensions
(in inches) are shown in Table I. These results should be compared with the experimental plot in Fig. 6 and the finite element
analysis results of Fig. 12. Additional discussion of this topic
can be found in reference 21. Although additional refinements
have been made to this model[40-45], the assumptions made with regard
to σ_s are phenomenologically incorrect and hence the technique is
not readily extendable to the defect characterization problem.

    Most of the analytical results associated with eddy current
NDT methods have been obtained for two geometries; a cylindrical
conductor surrounded by a concentric excitation coil and a circular
excitation coil above a conducting plate[28,46-50]. In general, the
approach has been to manipulate Maxwell's equations into a form
suitable for solution by standard partial differential equation
techniques such as Bessel functions, power series and Fourier
transforms.

    For example, Libby[28] makes use of magnetic vector potential,
$\bar{A}$ where

$$\bar{B} = \text{curl } \bar{A} \tag{9}$$

to obtain the partial differential equation

$$\frac{\partial^2 A_\theta}{\partial r^2} + \frac{1}{r}\frac{\partial A_\theta}{\partial r} - \frac{A_\theta}{r^2} + \sigma^2\mu\epsilon A_\theta - j\omega\mu\sigma A_\theta = 0 \tag{10}$$

describing the vector potential for the conducting cylinder and
encircling coil. An expression for A_θ is obtained in terms of

Table I. Slot Dimensions for Figs. 9 and 10

Slot	Width (Inches)	Depth (Inches)
1	0.125	0.10
2	0.125	0.15
3	0.125	0.20
4	0.125	0.25
5	0.125	0.30

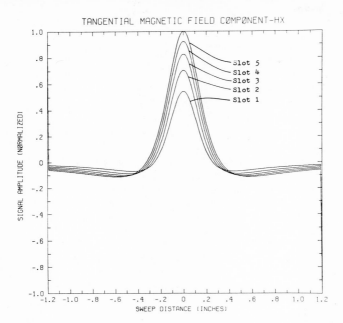

Fig. 9.   Dipole model predictions of residual H_x leakage field
profiles around rectangular slots in steel specimens.

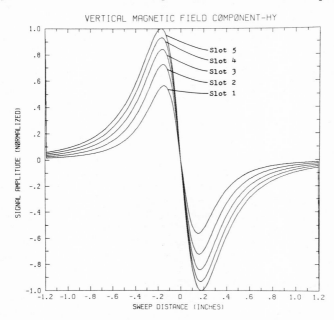

Fig. 10.   Dipole model predictions of residual H_y leakage field
profiles around rectangular slots in steel specimens.

Bessel functions and by making use of the identity

$$e_i = - \int_0^{2\pi} r \frac{\partial A_\theta}{\partial t} d\theta \qquad (11)$$

the coil voltage is deduced. In many cases, results are obtained in integral equation form which can be solved by numerical analysis techniques on the computer[51,52]. Attempts have also been made to include the effects of defects analytically by making use of dipole analogs and diffraction theory[53-55]. A major drawback of the analytical approach, as far as the defect characterization problem is concerned, relates to the inherent mathematical complexity of the problem and the difficulty associated with extending the limited results already obtained to realistic defects in realistic materials[56].

An approach to the field/defect interaction problem which does show promise of overcoming some of the drawbacks associated with analytical studies is the finite element method, a numerical technique originally developed for the study of electromagnetic fields in electrical machinery[57-64].

The nonlinear Poisson equations in two dimensions describing active, residual, and eddy current fields are respectively:

$$\frac{\partial}{\partial x} (\nu \frac{\partial A}{\partial x}) + \frac{\partial}{\partial y} (\nu \frac{\partial A}{\partial y}) = -J \qquad (12)$$

$$\frac{\partial}{\partial x} (\nu \frac{\partial A}{\partial x}) + \frac{\partial}{\partial y} (\nu \frac{\partial A}{\partial y}) = 0 \qquad (13)$$

$$\frac{\partial}{\partial x} (\nu \frac{\partial A}{\partial x}) + \frac{\partial}{\partial y} (\nu \frac{\partial A}{\partial y}) = j\sigma\omega A - J_s \qquad (14)$$

where ν is $1/\mu$, the material reluctivity, which must be modeled appropriately for the materials' region of operation as summarized in Fig. 1, and J_s, in the eddy current case, is the source current density. Unlike analytical techniques, the finite element method does not seek to solve equations 12, 13 and 14 directly but rather, by minimizing the corresponding energy functional over the discretized region of interest, gives values for the magnetic vector potential, point by point, which satisfy the describing equations.

Such techniques can be applied to electromagnetic NDT problems[65-70] and Figs. 11, 12 and 13 give typical results. Fig. 11 shows the mesh structure associated with a finite element study of the magnetic fields around the simple reluctance probe of Fig 3, together with a typical flux plot for the probe entering the crevice gap (support plate) region. By allowing the probe to

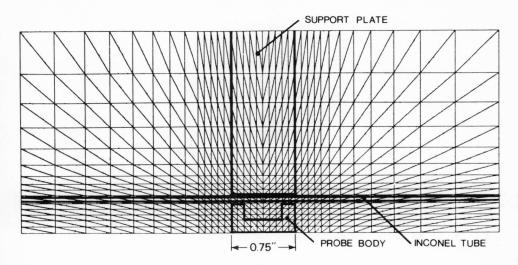

Fig. 11.   Finite element prediction of "active" fields associated
            with the variable reluctance probe shown in Fig. 3.
            a) Mesh structure.

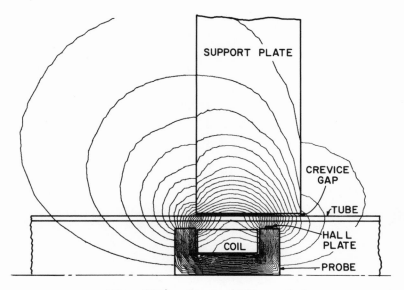

Fig. 11b). Typical flux plot.

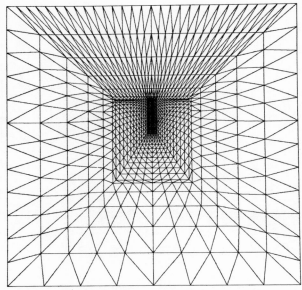

Fig. 12.   Finite element prediction of residual leakage field
profiles around a rectangular slot in a steel specimen.
a)   Mesh structure.

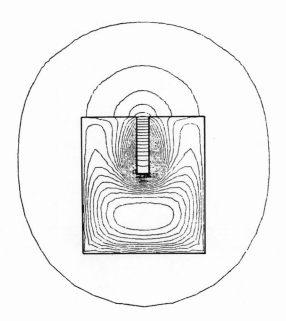

Fig. 12b). Typical flux plot.

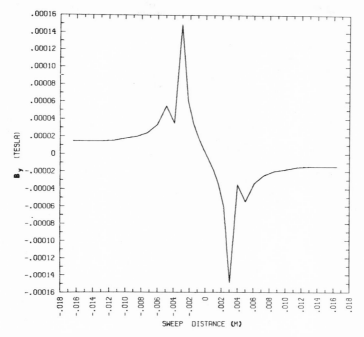

Fig. 12c).   Theoretical prediction of the slot residual leakage
             field profile.

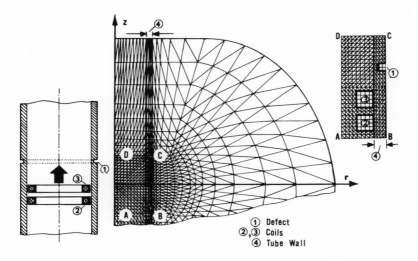

Fig. 13.   Finite element prediction of eddy   current impedance
           plane trajectories at 100kHz.   a)   Mesh structure.

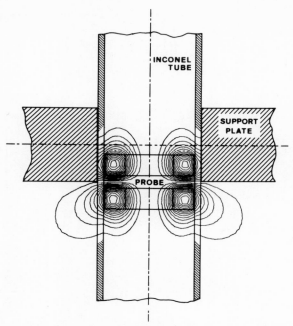

Fig. 13b).   Typical flux plot for a differential probe entering
the support plate region.

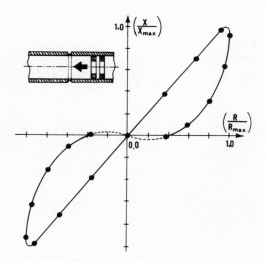

Fig. 13c).   Corresponding impedance plane trajectory for the
probe passing an O.D. axisymmetric slot.

increment through the support plate, a prediction can be made of the Hall plate output signal and this theoretical result is shown superimposed over the corresponding experimental data in Fig. 4.

Fig. 12a) shows the mesh used to predict the residual leakage flux around a rectangular slot in a steel bar and Fig. 12b) gives the corresponding leakage flux plot following a magnetization current of 150 amps. The magnetic characteristics used in this residual field study were similar to those shown in the first and second quadrants of Fig. 1. The finite element prediction of the leakage field profile is given in Fig. 12c) and should be compared to the experimental result of Fig. 6.

Fig. 13a) gives details of the mesh structure for an eddy current study related to the prediction of impedance plane trajectories for axisymmetric slots in Inconel tubing. A typical flux plot is given in Fig. 13b) for a differential eddy current probe at 100kHz together with the corresponding Lissajous pattern in Fig. 13c) which can be compared to the experimental result of Fig. 8.

CONCLUDING REMARKS

Electromagnetic methods of nondestructive testing have a history dating back almost a century. Without doubt, because of the inherent simplicity of these testing methods, they will remain an important industrial tool for many centuries to come. Equipment developments in the years ahead will most likely relate to the automation of testing procedures including defect characterization via appropriate signal processing techniques. Some advances can be expected in transducer development particularly with regard to miniaturization and the development of probes for reliably detecting the presence of even smaller defect sizes. An example of this is the recent production of micron-size Hall plates for high resolution magnetic field measurements[71,72].

On the theoretical front, significant strides will be made both with regard to a more fuller understanding of electromagnetic field/material interactions and the development of defect characterization schemes. Numerical analysis techniques will play a dominant role in modeling studies and with the advent of a new generation of computers we will see these modeling schemes extended to three dimensions[73-76]. In order to do this, the materials under test will have to be characterized with regard to conductivity and permeability to a much higher level of accuracy than presently available.

In addition to equipment and modeling developments, unforeseeable breakthroughs will also occur in the application of known electromagnetic phenomena to new testing situations and/or new

electromagnetic phenomena to existing testing situations. Pulsed
eddy current, multi-frequency eddy current, microwave and Barkhausen
phenomena will all receive increasing attention during the coming
years.

The purpose of this paper has been to give a somewhat sketchy
overview of electromagnetic NDT methods. The serious reader will
find additional "food for thought" in the following extensive
reference section.

REFERENCES

1.   D. E. Hughes, Induction Balance and Experimental Researches
     Therewith, Phil. Mag., 8:50 (1879).
2.   W. Lord and D. J. Oswald, Leakage Field Methods of Defect
     Detection, Int. J. NDT, 4:249 (1972).
3.   R. C. McMaster, "Nondestructive Testing Handbook," Vol. II,
     ASNT, Columbus (1959).
4.   W. J. McGonnagle, "Nondestructive Testing," Gordon and Breach,
     New York (1961).
5.   J. Lankford and P. H. Francis, Magnetic Field Perturbation Due
     to Metallurgical Defects, Int. J. NDT, 3:77 (1971).
6.   T. Shiraiwa, T. Hiroshima and S. Morishima, An Automatic
     Magnetic Inspection Method Using Magnetoresistive Elements
     and its Application, Mat. Eval., 31:90 (1973).
7.   K. C. Bakhaus, Magnetic Field Measurements as Related to
     Industrial NDT Applications, Mat. Eval., 33:232 (1975).
8.   P. A. Khalileev and P. A. Grigorev, Method of Testing the Con-
     dition of Underground Pipes in Main Pipelines, Defektoskopiya,
     4:79 (1974).
9.   W. Stumm, Magnetic Stray Flux Measurement for Testing Welded
     Tubes On-Line, Non-Destructive Testing, Feb:3 (1976).
10. F. Förster, "The Automatic Detection of Flaws and Marking of
     Flaws by the Magnetographic Method," Förster Inst. Reutlingen
     (1964).
11. D. E. Lorenzi, et. al., Classifying Seam Depths in Steel
     Billets by the Magnetic Tape Method, Mat. Eval., 27:238 (1969).
12. Y. B. Feshchenko, The Prospects of Magnetographic Defectoscopy
     of Rolled Iron, Defektoskopiya, 2:5 (1971).
13. M. S. Gieskieng, "Method and Apparatus for the Detection of
     Cracks and Flaws in Rail Wheels and the Like by Sliding a Pre-
     recorded Magnetic Medium over the Test Piece," United States
     Patent 3,820,016, June 25, 1974.
14. W. Lord, Magnetic Flux Leakage for Measurement of Crevice Gap
     Clearance and Tube Support Plate Inspection, in: EPRI Special
     Report - Nondestructive Evaluation Program: Progress in 1979,
     EPRI, Palo Alto (1979).
15. N. Davis, Magnetic Flux Analysis Techniques, in: "Research
     Techniques in Nondestructive Testing," R. S. Sharpe, ed.,
     Academic Press, New York (1973).

16. V. A. Burganova, et. al., Electromagnetic Testing of the Microstructure and Mechanical Properties of Cold-Deformed Tubes, Defektoskopiya, 2:46 (1972).

17. S. Titto, et. al., Non-destructive Magnetic Measurement of Steel Grain Size, Non-Destructive Testing, June:117 (1976).

18. V. K. Barsukov, et. al., Inspection of Ferromagnetic Material Parameters by the Magnetic Noise Method, Defektoskopiya, 6:117 (1973).

19. G. A. Matzkanin, et. al., "The Barkhausen Effect and its Applications to Nondestructive Evaluation," NTIAC-79-2, SRI, San Antonio, (1979).

20. N. S. Savorovskii, et. al., Contactless System for Transverse Magnetization of Tubes, Defektoskopiya, 2:23 (1970).

21. W. Lord and D. J. Oswald, The Generated Reaction Field Method of Detecting Defects in Steel Bars, Mat. Eval., 29:21 (1971).

22. F. Förster, Non-destructive Inspection of Tubing and Round Billets by means of Leakage Flux Probes, Brit. J. NDT, 19:26 (1977).

23. R. E. Beissner, "An Investigation of Flux Density Determinations," AFML Final Report TR-76-236, SRI, San Antonio, (1976).

24. I. I. Kifer and I. B. Semenouskaya, New Magnetic Particle Methods of Inspection, Defektoskopiya, 2:41 (1972).

25. W. Lord, et. al., Residual and Active Leakage Fields Around Defects in Ferromagnetic Materials, Mat. Eval., 36:47 (1978).

26. G. M. Massa, Finding the Optimum Conditions for Weld Testing by Magnetic Particles, Non-Destructive Testing, Feb:16 (1976).

27. D. A. Shturkin, Magnetization of Components with Difficult Shapes for Flaw Detection Purposes, Defektoskopiya, 4:63 (1976).

28. H. L. Libby, "Introduction to Electromagnetic Nondestructive Test Methods," Wiley-Interscience, New York (1971).

29. R. Hochschild, Electromagnetic Methods of Testing Metals, in: "Progress in Non-Destructive Testing," E. G. Stanford, et. al., Ed., Macmillan Company, Inc., New York (1959).

30. J. R. Wait, Review of Electromagnetic Methods in Nondestructive Testing of Wire Ropes, Proc. IEEE, 67:892 (1979).

31. V. S. Cecco and C. R. Box, Eddy Current in-situ Inspection of Ferromagnetic Monel Tubes, Mat. Eval., 33:1 (1975).

32. C. V. Dodd and W. A. Simpson, Thickness Measurements Using Eddy Current Techniques, Mat. Eval., 31:73 (1973).

33. C. V. Dodd, et. al., "Eddy Current Evaluation of Nuclear Control Rods," Mat. Eval., 32:93 (1974).

34. H. L. Libby, Eddy Current Test for Tubing Flaws in Support Regions, in: "Research Techniques in Nondestructive Testing," R. S. Sharpe, Ed., Academic Press, New York (1973).

35. A. J. Bahr, Quantitative Measurement of Crack Parameters using Microwave Eddy Current Technqiues, in: Proceedings of the ARPA/AFML Review of Progress in Quantitative NDE, AFML-TR-78-205 (1979).

36. D. S. Dean and L. A. Kerridge, Microwave Techniques, "Research Techniques in Nondestructive Testing," R. S. Sharpe, Ed. Academic Press, New York (1970).
37. D. L. Waidelich, Pulsed Eddy Currents, ibid.
38. N. N. Zatsepin and V. E. Shcherbinin, Calculation of the Magnetostatic Field of Surface Defects, I. Field Topography of Defect Models, Defektoskopiya, 5:50 (1966).
39. V. E. Shcherbinin and N. N. Zatsepin, Calculation of the Magnetostatic Field of Surface Defects, II. Experimental Verification of the Principal Theoretical Relationships, Defektoskopiya, 5:59 (1966).
40. V. E. Shcherbinin and A. I. Pashagin, Influence of the Extension of a Defect on the Magnitude of Its Magnetic Field, Defektoskopiya, 4:74 (1972).
41. I. A. Novikova and N. V. Miroshin, Investigation of the Fields of Artificial Open Flaws in a Uniform Constant Magnetic Field, Defektoskopiya, 4:95 (1973).
42. G. A. Burtsev and E. E. Fedorishcheva, Simple Approximation for the Magnetostatic Fields of Surface Defects and Inhomogeneities, Defektoskopiya, 2:111 (1974).
43. V. E. Shcherbinin and A. I. Pashagin, Polarization of Cracks in Nonuniformly Magnetized Parts, Defektoskopiya, 3:17 (1974).
44. V. E. Shcherbinin and A. I. Pashagin, On the Volume Polarization of Cracks, Defektoskopiya, 4:106 (1974).
45. B. I. Kolodii and A. Y. Teterko, Determination of the Transverse Magnetostatic Field of a Cylinder with an Eccentric Cylindrical Inclusion, Defektoskopiya, 3:44 (1976).
46. R. Hochshild, Electromagnetic Methods of Testing Metals, in: "Progress in Non-Destructive Testing," E. G. Stanford, et. al., Ed., Macmillan Company, Inc., New York (1959).
47. P. Graneau and S. A. Swann, Electromagnetic Fault Detection in Non-ferrous Pipes, J. Electronics and Control, 8:127 (1960).
48. J. Vine, Impedance of a Coil Placed Near to a Conducting Sheet, J. Electronics and Control, 16:569 (1964).
49. D.H.S. Cheng, The Reflected Impedance of a Circular Coil in the Proximity of a Semi-Infinite Medium, IEEE Trans. IM, 14:107 (1965).
50. C. V. Dodd, "Solutions to Electromagnetic Induction Problems," Ph.D. Dissertation, University of Tennessee (1967).
51. C. V. Dodd, W. E. Deeds and J. W. Luquire, Integral Solutions to Some Eddy Current Problems, Int. J. NDT, 1:29 (1969).
52. C. V. Dodd, The Use of Computer Modeling for Eddy Current Inspection, in: "Research Techniques in Nondestructive Testing," R. S. Sharpe, Ed., Academic Press, New York (1977).
53. M. Burrows, "Theory of Eddy Current Flaw Detection," Ph.D. Thesis, University of Michigan (1964).
54. D. A. Hill and J. R. Wait, Scattering by a Slender Void in a Homogeneous Conducting Wire Rope, Appl. Phys., 16:391 (1978).

55. A. H. Kahn, R. Spal and A. Feldman, Eddy-current Losses due
     to a Surface Crack in Conducting Material, J. Appl. Phys.,
     48:4454 (1977).
56. W. Lord and R. Palanisamy, Development of Theoretical Models
     for NDT Eddy Current Phenomena, to appear in the ASTM Journal.
57. A. M. Winslow, Numerical Solution of the Quasilinear Poisson
     Equation in a Nonuniform Triangle Mesh, J. Comp Phys., 2:149
     (1967).
58. P. Silvester and M.V.K. Chari, Finite Element Solution of
     Saturable Magnetic Field Problems, IEEE Trans. PAS., 89:1642
     (1970).
59. O. W. Anderson, Transformer Leakage Flux Program Based on the
     Finite Element Method, IEEE Trans. PAS., 92:682 (1973).
60. M.V.K. Chari, Finite Element Solution of the Eddy Current
     Problem in Magnetic Structures, IEEE Trans. PAS., 93:62 (1974).
61. J. H. Hwang and W. Lord, Finite Element Analysis of the Mag-
     netic Field Distribution Inside a Rotating Ferromagnetic Bar,
     IEEE Trans. MAG., 10:1113 (1974).
62. T. Sato, et. al., Calculation of Magnetic Field Taking into
     Account Eddy Current and Nonlinear Magnetism, Elec. Eng. in
     Japan, 96:96 (1976).
63. M.V.K. Chari and Z. J. Csendes, Finite Element Analysis of
     the Skin Effect in Current Carrying Conductors, IEEE Trans.
     MAG., 13:1125 (1977).
64. N. A. Demerdash and T. W. Nehl, An Evaluation of the Methods
     of Finite Elements and Finite Differences in the Solution of
     Nonlinear Electromagnetic Fields in Electrical Machines,
     IEEE Trans. PAS., 98:74 (1979).
65. J. Donea, et. al., Finite Elements in the Solution of Electro-
     magnetic Induction Problems, Int. J. Numerical Methods in Eng.,
     8:359 (1974).
66. W. Lord and J. H. Hwang, Finite Element Modeling of Magnetic
     Field/Defect Interactions, ASTM J. Testing and Eval., 3:21
     (1975).
67. W. Lord and J. H. Hwang, Defect Characterization from Magnetic
     Leakage Fields, Brit. J. NDT, 19:14 (1977).
68. T. G. Kincaid and M.V.K. Chari, The Application of Finite
     Element Method Analysis to Eddy Current NDT, Proc. of the
     ARPA/AFML Review of Progress in Quantitative NDE, LaJolla
     (1978).
69. R. Palanisamy and W. Lord, Finite Element Analysis of Axisym-
     metric Geometries in Quantitative NDE, Proc. of the ARPA/AFML
     Review of Progress in Quantitative NDE, LaJolla (1979).
70. R. Palanisamy and W. Lord, Finite Element Modeling of Electro-
     magnetic NDT Phenomena, IEEE Trans. MAG., 15:1479 (1979).
71. H. T. Minden and M. F. Leonard, A Micron-size Hall Probe for
     Precision Magnetic Field Mapping, J. Appl. Phys., 50:2945
     (1979).

72. A. W. Baird, et. al., High Resolution Field Measurements Near
     Ferrite Recording Heads, IEEE Trans. MAG., 15:1631 (1979).
73. O. C. Zienkiewicz, et. al., Three-Dimensional Magnetic Field
     Determination Using Scalar Potential - A Finite Element Solu-
     tion, IEEE Trans. MAG., 13:1649 (1977).
74. E. Guancial and S. Das Gupta, Three-Dimensional Finite Element
     Problems, IEEE Trans. MAG., 13:1012 (1977).
75. C. J. Carpenter, Comparison of Alternative Formulations of
     3-Dimensional Magnetic Field and Eddy Current Problems at
     Power Frequencies, Proc. IEE., 124:1026 (1977).
76. T. W. Preston and A.B.J. Reece, Finite Element Solution of
     3-Dimensional Eddy Current Problems in Electrical Machines,
     Proc. of the COMPUMAG Conference, Grenoble (1978).

ACKNOWLEDGMENTS

    Much of the work described in this paper could not have been
completed without the financial assistance of the Colorado Energy
Research Institute, the Army Research Office and the Electric
Power Research Institute. In addition, I have been fortunate to
have worked with excellent students over the years on electromagnetic NDT problems. Many of the modeling results reported in this
paper were due to the efforts of J. H. Hwang, R. Palanisamy,
J. M. Bridges, T. J. McCauley, W. C. Yen and S. R. Satish.

STIFFNESS CHANGE AS A NONDESTRUCTIVE DAMAGE MEASUREMENT

T. Kevin O'Brien

Structures Laboratory
U.S. Army Research and Technology
   Laboratories (AVRADCOM)
NASA Langley Research Center
Hampton, VA   23665

ABSTRACT

In laboratory specimens, damage growth is generally accompanied by stiffness change. Hence, damage growth can be monitored
indirectly through stiffness. Furthermore, stiffness measurements
are both nondestructive and easier to apply than most NDI methods
that measure damage directly. This paper highlights laboratory
applications of stiffness measurement for indirect assessment of
damage growth and, in some special cases, for failure prediction.
Examples cited are buckling of compressively loaded cylindrical
shells, slow stable crack growth in tension-loaded notched coupons,
fatigue-crack growth in adhesively bonded materials, and fatigue-
damage growth in composite materials.

INTRODUCTION

As damage accumulates in laboratory specimens, their stiffness
changes. (Herein, the term stiffness can typically be associated
with the slope of a load-displacement record or a stress-strain
curve. More precise definitions are given, when they appear necessary, in discussions of specific applications.) Stiffness change
can be used as a quantitative record of damage growth. Hence,
damage growth can be monitored indirectly through stiffness. Furthermore, nondestructive stiffness measurements are easier to
accomplish than most direct measurements of damage obtained by
other NDI methods. In addition, stiffness change is sometimes the
critical parameter that determines failure. One example is a

101

stiffness-critical phenomenon, buckling. Stiffness change has also
been used to predict fatigue failure for specific composite materials.

This paper highlights laboratory applications of stiffness
measurement for indirect assessment of damage growth and, in some
special cases, for failure prediction. Examples cited are buckling
of compressively loaded cylindrical shells, slow stable crack growth
in tension-loaded notched coupons, fatigue-crack growth in adhesively bonded materials, and fatigue-damage growth in composite
materials.

BUCKLING OF COMPRESSIVELY LOADED CYLINDERS

Perhaps the most obvious application of stiffness measurement
is failure prediction in stiffness critical applications. For
example, the buckling of compressively loaded columns has been predicted using transverse normal stiffness measurements.[1] The transverse normal stiffness decreased linearly with increasing axial
compressive load, and finally vanished at the buckling load. Hence,
transverse normal stiffness measurements, taken at small axial compressive loads, can be extrapolated to predict the column buckling
load.

The same technique has been applied to compressively loaded
cylindrical shells both to locate the initial buckling site and to
predict the buckling load.[2] The shells had various stiffener types
and cross-sectional shapes. The test procedure was as follows.
Small forces were applied, one at a time, normal to the shell surface at circumferential stations by a local static stiffness probe.
The circumferential distribution of transverse normal stiffness for
a longitudinally-stiffened circular shell is shown in figure 1.
Distributions generated at axial compressive loads of zero and
1780 N were quite similar. The region of "least" stiffness was
noted and the stiffness variation in that region was plotted for
several levels of applied axial compressive load (fig. 2). The rate
of decrease in transverse normal stiffness with increasing axial
load was greatest at stations 03 and 04. A linear extrapolation of
these data to zero transverse normal stiffness was used to predict
the buckling load for the shell. The following table summarizes
predicted and observed buckling loads for the various shell configurations tested. In none of the tests was it necessary to apply
axial loads greater than one-third of the buckling load. Predictions were accurate for five of the six configurations tested.
Hence, test results on several conventional shell configurations
showed that excellent predictions of the buckling load can be made
from transverse normal stiffness data obtained at relatively low
compressive loads.

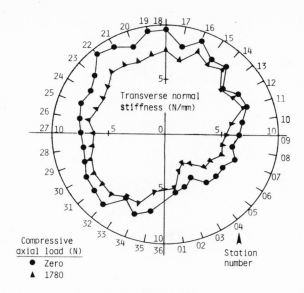

Fig. 1. Transverse normal stiffness profile at the mid-length of a longitudinally stiffened circular cylindrical shell.

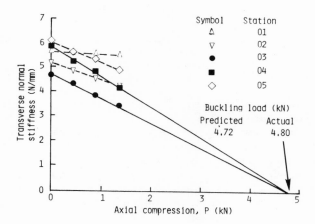

Fig. 2. Transverse normal stiffness-axial compression plots for the longitudinally stiffened circular shell at region of least stiffness.

Cylindrical shell	Transverse normal force (N)	Buckling load (N)		Error %
		Predicted	Actual	
Unstiffened circular	1.96	6183	6116	+1.1
Stringer stiffened circular	4.9	4715	4804	−1.9
Ring and stringer stiffened circular	4.9	8674	8896	−2.5
Unstiffened elliptic	0.74	1601	1650	−3.0
Unstiffened elliptic with cutout	0.74	800	850	−5.8
Spiral stiffened circular	5.9	20907	15124	+38.2

SLOW STABLE CRACK GROWTH

One way to predict failure of cracked tensile coupons that exhibit some slow stable crack growth involves the compliance (reciprocal of stiffness) technique of linear elastic fracture mechanics. The compliance technique is based on the relation between strain-energy-release rate and the rate of change of compliance with crack growth.[3,4] For an elastic body, the strain-energy-release rate, G, is the difference between the rate of work done, dW/dA, and the rate at which elastic strain energy is stored, dU/dA, as the crack surface area A increases, i.e.,

$$G = \frac{dW}{dA} - \frac{dU}{dA} \qquad (1)$$

The compliance of the specimen increases as the crack grows. For example, figure 3 illustrates the increased compliance of a flat plate, having thickness t, with a centrally located through-crack growing under an applied tension load. Substituting into equation (1) expressions for the work done on the plate by a constant applied force[*], P, resulting in an axial deflection, u, at the point of load application

$$W = P \cdot u$$

and the strain energy stored in the body

$$U = \frac{1}{2} P \cdot u$$

along with Hooke's law, u = CP, where C is compliance, yields

$$G = \frac{P^2}{2t} \frac{dC}{da} \tag{2}$$

Although only a special configuration was considered here, a more rigorous analysis[5] shows equation (2) to be quite general. Therefore, the strain-energy-release rate, G, is directly related to the rate of change of compliance with crack length, dC/da. Furthermore, simple equations have been derived[4] to directly relate the stress-intensity factor, K, to the strain-energy-release rate, G. Hence, K is also related to the rate of change of compliance with crack length. (A review of the application of compliance techniques to a variety of cracked-specimen configurations and types of loading is given by Okamura, et al.[6])

In many cracked coupons monotonically loaded in tension, the crack extends stably and slowly as the load is increased. The slow stable growth results from energy absorbing mechanisms, other than crack extension, occurring at the crack tip. For example, in metal sheets the energy absorbing mechanism is plastic yielding, whereas in random-fiber epoxy composites it is fiber-matrix debonding accompanied by matrix crazing. In either case, elastic analyses for stress-intensity factors do not account for these other mechanisms. However, crack-growth-resistance curves (R-curves) can be generated using compliance measurements.[7] These R-curves can be used with calculated elastic stress-intensity factors to determine failure loads.

The following example illustrates a compliance technique used to generate R-curves and, through them, determine failure loads of notched, random-fiber, glass/epoxy composites.[8] Single-edge-notch tension specimens were tested. Crack-mouth opening displacements, δ, were measured with an extensometer mounted on two thin strips

---

[*]Equation (2) could also be derived assuming the other extreme of a constant applied displacement. Hence, the energy released during incremental crack extension is independent of loading configuration.

bonded to the specimen. Compliance, C, was calculated from the
initial slope of the graph of crack-mouth opening displacement, δ,
as a function of applied load, P. A compliance calibration curve
(fig. 4) was generated by plotting calculated compliance as a function of notch size for specimens with various sizes of machined
notches. Then, during a crack propagation test (fig. 5), the compliance (reciprocal of secant modulus) at prescribed load levels was
measured. The corresponding effective crack length[†] was determined
from the compliance calibration curve (fig. 4). Next, the applied
load and effective crack length were substituted into a previously
developed expression[9] for the stress-intensity factor, K, for a
single-edge-notch tensile specimen

$$K = Y \frac{P\sqrt{a}}{tw} \qquad\qquad\qquad (3)$$

where

Y   is a compliance calibration factor

a   is the effective crack length corresponding to load P

t   is the specimen thickness

w   is the specimen width

Stress-intensity factors were plotted as a function of crack extension, $a - a_0$, where a_0 is initial crack length. Figure 6 represents the crack-growth resistance, K_R, as a function of crack
extension for a random-fiber, glass/epoxy composite. Data points
generated from specimens with three different initial crack lengths
were plotted to illustrate that the R-curve is independent of initial crack length. Finally, the stress-intensity factor, K, was
calculated from equation (3) for several applied loads and plotted
as a function of crack length (fig. 7). These K-curves are commonly
referred to as crack-driving-force curves. Next, the crack-growth-
resistance curve was superimposed on figure 7. The crack-driving-
force curve generated for an applied load of 600 lbs was tangent to
the crack-growth-resistance curve. The point of tangency is the
instability point.[10] The value of K_R at the instability point is
the critical stress-intensity factor for unstable crack extension.

[†]The term "effective crack length" was used because damage produced
during crack growth is not identical to the machined notches.
However, the relationship between crack length and machined notch
size is defined through comparative compliance measurements. The
compliance associated with the growing crack and material damage
in front of the crack tip is equated with the compliance for a
particular notch size, to define the effective crack length.

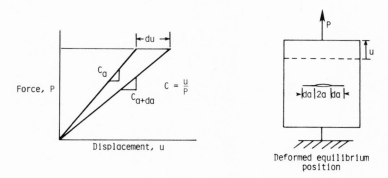

Fig. 3.   Schematic of compliance increase with crack extension.

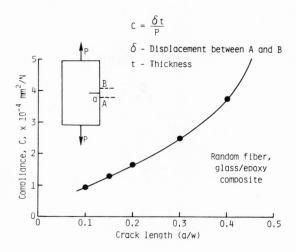

Fig. 4.   Compliance calibration curve for single-edge notched,
           random-fiber, glass/epoxy composite.

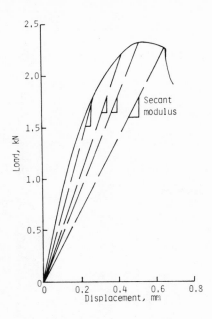

Fig. 5.   Compliance determination at various load levels during
          crack propagation test.

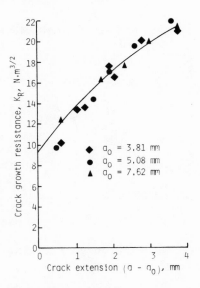

Fig. 6.   R-curve for random-fiber, glass/epoxy composite.

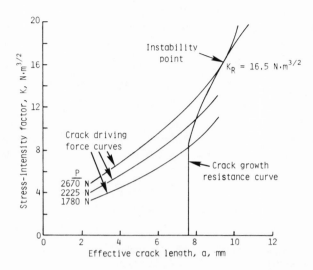

Fig. 7. Determination of crack instability point from the R-curve.

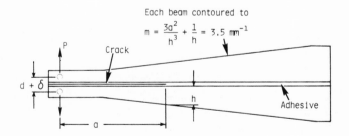

Fig. 8. Tapered Double Cantilever Beam specimen.

FATIGUE–CRACK GROWTH IN ADHESIVELY BONDED MATERIALS

Fatigue-crack growth in the adhesive layer of a bonded specimen can be estimated from compliance measurements. To do this easily, the specimen is constructed so that its compliance is a linear function of crack length. Then the fatigue-crack growth rate, da/dN, can be estimated from the rate of increase in compliance, dC/dN. Furthermore, since dC/da is constant, the strain-energy-release rate, G, varies only with the applied load and is independent of crack length. Hence, fatigue-crack growth rates, da/dN, and strain-energy-release rates, G, can be generated without directly measuring crack growth.

The specimen shown in figure 8 is one that is widely used. It is known as the Tapered Double Cantilever Beam (TDCB) specimen.[11] The TDCB specimen is tapered such that simple beam theory yields

$$\frac{dC}{da} = \frac{8}{Et}\, m \tag{4}$$

where

   C   is the specimen compliance calculated from the ratio of crack-mouth opening displacement, δ, measured at the load point, to the applied load, P

   a   is crack length measured from the point of load application

   E   is Young's modulus of the tapered beam

   t   is the specimen thickness

and

$$m = \left(\frac{3a^2}{h^3} + \frac{1}{h}\right) \tag{5}$$

where h is the height of the tapered beam. In practice, a constant value of m represents a contour that can be approximated by a straight line over relatively long distances for specimen construction purposes. For large constant values of m, the expression for G resulting from substituting equation (4) into equation (2) varies significantly only with the applied load.

The TDCB has been used to develop fatigue growth laws for cracks growing in the adhesive layer between bonded metal structures.[12] The TDCB specimen is useful for generating baseline data of the form $da/dN = c(\Delta G)^b$ for two reasons. First, the value of G can be obtained from the applied load without knowing the

crack length. Second, because compliance varies linearly with crack
length, the crack length is easily estimated from displacements measured at the load point. This eliminates the necessity for either
visually monitoring the crack growth (which is difficult for adhe-
sives) or using other more exotic and expensive crack-length monitoring methods. Figure 9 shows examples of the power-law correlation between fatigue-crack growth rate, da/dN, and the Mode I
strain-energy-release rate range, ΔG_I, during constant-amplitude
cyclic loading of 7075-T6 aluminum tapered beams bonded with AF-55S
adhesive film.

FATIGUE-DAMAGE GROWTH IN COMPOSITE MATERIALS

    The stiffness of many fibrous composite laminates degrades
noticeably early in their fatigue life.[13-19] Consequently, for
specific composite materials and configurations, researchers have
exploited measured stiffness changes to explain or predict fatigue
behavior. For example, stiffness changes were used to determine
the effects of frequency on fatigue-damage accumulation in notched
$[0/\pm45/0]_s$ boron/epoxy and boron/aluminum laminates.[15] To illustrate the frequency dependence of damage growth, S-N curves were
plotted. During the fatigue tests, dynamic compliance (reciprocal
of dynamic stiffness, defined as the slope of the secant connecting
the maximum and minimum points on the hysteretic cyclic stress-
strain curve) was measured by a 25.4-mm (1.0-in.) extensometer
across the 6.35-mm (0.25-in.) diameter center hole. Fatigue life
was defined as the number of cycles required to produce a specific
(18-20 percent) dynamic stiffness change (figs. 10 and 11). It was
found that at all cyclic load amplitudes, fewer cycles were required
at lower frequencies to generate damage resulting in identical
stiffness changes. Hence, if low frequencies are expected in
service loading of a component, then low-frequency tests should be
run to accurately reflect the fatigue behavior of the component.

    In two other applications which involved studies on boron/
aluminum laminates, stiffness changes correlated linearly with
residual strength. In one study[16] the elastic static tangent modulus (measured during unloading) of unnotched $[0/\pm45/90/0/\pm45/\overline{90}]_s$
laminates correlated with the residual strength measured after
2-million fatigue cycles (fig. 12). In another study[17] the dynamic
stiffness (secant modulus) measured over a 25.4-mm (1.0-in.) gage
length in a $[0/\pm45/0]_s$ laminate containing a 6.35-mm (0.25-in.)
diameter center hole correlated with the dynamic fracture stress
after 1.5-million cycles (fig. 13). Such correlations indicate
that measured stiffness change may provide a valid estimate of
residual strength in boron/aluminum composites.

    In yet another application, stiffness change was used to pre-
dict real-time fatigue failure of unnotched, quasi-isotropic,

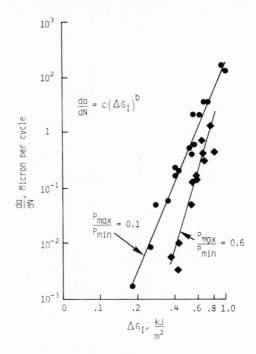

Fig. 9.   Fatigue crack growth rate as a function of ΔG_I for
          an adhesively bonded metal.

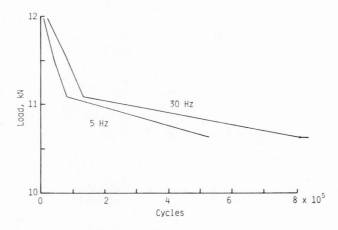

Fig. 10.   S-N curve based on constant stiffness change for
           notched boron/epoxy laminate.

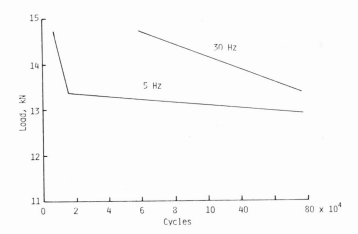

Fig. 11.   S-N curve based on constant stiffness change for
          notched boron/aluminum laminate.

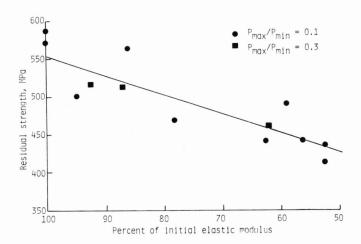

Fig. 12.   Residual strength correlation with change in elastic
          unloading modulus for $[0/\pm45/90/0/\pm45/\overline{90}]_s$ boron/
          aluminum laminates.

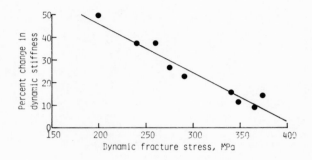

Fig. 13. Correlation of dynamic stiffness change with dynamic
          fracture stress for notched $[0/\pm45/0]_s$ boron/aluminum
          laminates.

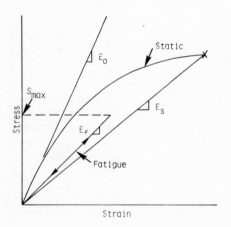

Fig. 14. Schematic diagram of secant modulus change.

glass/epoxy laminates.[18] In this application, observed stiffness
changes were compared with secant modulus degradation measured in
a static tensile strength test on identical laminates (fig. 14).
When the static modulus measured during fatigue, E_f, degraded from
its initial tangent modulus, E_o, to the secant modulus measured at
failure in a static ultimate strength test, E_s, then fatigue failure was predicted. Real-time fatigue failure was predicted using
this secant modulus criterion for load-controlled, constant-
amplitude, tension-tension cyclic loading of $[0/\pm45/90]_s$ glass/epoxy
laminates (fig. 15). The secant modulus criterion also successfully
predicted fatigue failure in $[0/90/\pm45]_s$ boron/epoxy laminates subjected to blocks of increasingly severe, constant-amplitude cyclic
loads.[19] However, it could not successfully predict fatigue failure
in $[0/90]_s$ boron/epoxy laminates under similar loading conditions.
Hence, the applicability of the secant modulus criterion for different materials, loading conditions, and stacking sequences is
questionable.

An initial part of an ongoing effort at VPI&SU to develop a
stiffness-based damage model for composite materials was the development of a technique for measuring the longitudinal, transverse,
and shear moduli on the same composite specimen.[19,20] The technique
has been used[19] to measure stiffness changes in unnotched, boron/
epoxy laminates subjected to blocks of increasingly severe constant-
amplitude cyclic loading. Figure 16 shows the relative degradations
of the longitudinal, transverse, and shear moduli for $[0/90]_s$ and
$[0/90/\pm45]_s$ laminates after 50,000 load cycles. The maximum cyclic
strain amplitude ranged from 3300 to 3500 $\mu m/m$ for $[0/90/\pm45]_s$ laminates and from 2800 to 3300 $\mu m/m$ for the $[0/90]_s$ laminates. Measurements like these should be useful in verifying the model when
it is completed.

Investigators at the U.S. Army Structures Laboratory at NASA's
Langley Research Center are using stiffness measurements in the
development of a generic fatigue analysis for arbitrary composite
laminates. The growth of both in-plane and interlaminar fatigue
damage is being characterized for inclusion in the model.

One technique for characterizing the growth of in-plane
fatigue damage in the off-axis plies of complex laminates has been
developed.[21] The relationship between measured stiffness reductions
and constant-amplitude fatigue cycles in simple, unnotched, angle-
ply laminates is assumed to have a Weibull distribution. Figure 17
shows a least-squares fit of a Weibull function to modulus reductions in $[\pm45]_{2s}$ boron/epoxy laminates. Changes in measured axial
deflections have been converted to reductions in the in-plane shear
modulus, G_{12}, for three different maximum cyclic loads corresponding to three different levels of in-plane shear stress, τ_{12}.
Each curve has an associated equation of the form

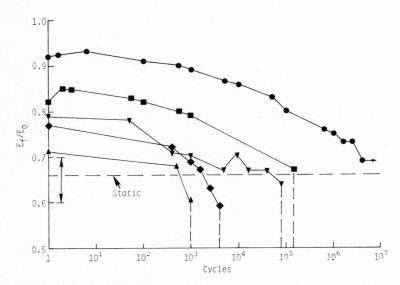

Fig. 15.   Fatigue failure prediction of $[0/\pm45/90]_s$ glass/epoxy
           laminates from secant modulus criterion.

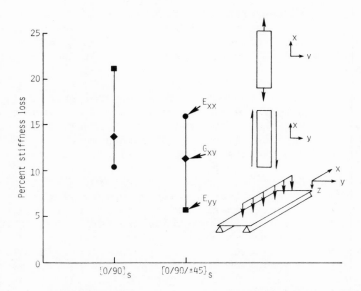

Fig. 16.   Stiffness reductions in boron/epoxy laminates after
           50,000 cycles.

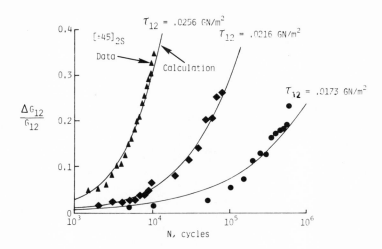

Fig. 17.   Weibull fit of in-plane shear modulus degradations with
           load cycles for $[\pm 45]_{2s}$ boron/epoxy laminates.

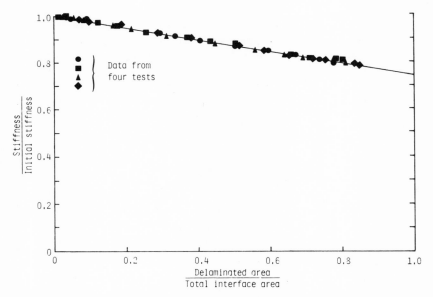

Fig. 18.   Linear stiffness degradation with delamination size in
           $[\pm 30/\pm 30/90/\overline{90}]_s$ graphite/epoxy laminates.

$$\frac{\Delta G_{12}}{\text{Initial } G_{12}} = 1 - e^{(N/\beta)^{\alpha}} \tag{6}$$

The parameters α and β, determined from the least-squares fit of equation (6), are functions of material properties and the maximum cyclic stress. Once enough baseline tests have been conducted to evaluate the parameters α and β, then equation (6) can be used, with an appropriate stress analysis, to simulate 45°-ply damage growth in arbitrary laminates subjected to complex load histories.

To characterize interlaminar damage growth, a delamination growth law of the form $da/dN = cG^b$ has been developed[22] where da/dN is the rate of growth of the delamination with applied fatigue cycles, G is the total strain-energy-release rate, and c and b are empirical parameters determined from a least-squares fit to a power law. The law was developed from strain-controlled, constant-amplitude, tension-tension fatigue tests on unnotched $[\pm30/\pm30/90/\overline{90}]_s$ graphite/epoxy coupons designed to delaminate at the straight edge. The stiffnesses of these specimens, measured with a linear variable differential transformer (LVDT) over a large gage length, degraded linearly with delamination size (fig. 18). Hence, delamination growth rates, da/dN, were estimated from stiffness measurements instead of being measured from the dye-penetrant-enhanced X-ray photographs required to directly reveal the delamination front (fig. 19). Furthermore, the strain-energy-release rate, G, was independent of delamination size and could be calculated from the maximum nominal cyclic strain. Therefore, a power-law fit could be performed using data generated from nominal strain and load measurements alone. Figure 20 compares the curve fits of da/dN as a function of G based on data obtained by a direct measurement technique and the stiffness technique. The power law fit the data well for both measurements of da/dN, with the law generated from stiffness measurements being the more conservative. Because both da/dN and G can be calculated without directly measuring the delamination size, the stiffness technique is a prime candidate for determining how the empirical parameters c and b are affected by different load histories, frequencies, temperatures, environments, etc. Once the parameters c and b are determined, the power law could be applied to more complicated configurations. Then, strain-energy-release rates calculated from an appropriate stress analysis could be used with the c and b parameters measured from straight-edge-delamination baseline tests to determine delamination growth rates.

CONCLUDING REMARKS

Laboratory applications of stiffness measurement for indirect assessment of damage growth and, in some special cases, real-time

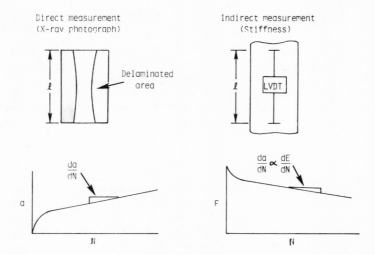

Fig. 19.   Delamination growth rate measurements.

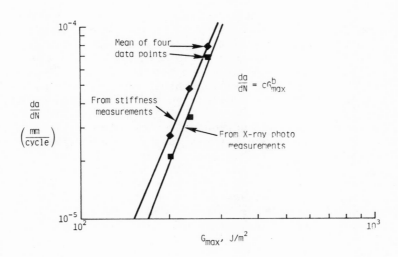

Fig. 20.   Delamination growth law using various growth
           rate measurements.

failure prediction were discussed. Examples were cited for buckling
of compressively loaded cylindrical shells, slow stable crack growth
in tension-loaded notched coupons, fatigue-crack growth in adhesively bonded materials, and fatigue-damage growth in composite
materials.

Stiffness measurement has been shown to provide a useful,
indirect assessment of damage growth in laboratory specimens for a
variety of materials, configurations, and loadings. In addition,
stiffness change can be used to predict buckling of columns and
cylindrical shells and fatigue failure of some specific composite
materials.

REFERENCES

1. W. H. Horton, J. I. Craig, and D. E. Struble, A Simple Practi-
    cal Method for the Experimental Determination of the End
    Fixity of a Column, Proceedings of the Eighth International
    Symposium on Space Technology and Science, Tokyo (1969).
2. W. H. Horton, E. M. Nassar, and M. K. Singhal, Determination
    of the Critical Loads of Shells by Non-Destructive Methods,
    Experimental Mech., 17(4):154 (1977).
3. G. R. Irwin and J. A. Kies, Critical Energy Rate Analysis of
    Fracture Strength, Welding Journal Research Supplement,
    33:193S (1954).
4. G. R. Irwin, Fracture, Handbuch Der Physik, 6:551 (1958).
5. D. Broek, "Elementary Engineering Fracture Mechanics,"
    Sijthoff and Noordhoff, Alphen · Aan Den Rijn (1978).
6. H. Okamura, K. Watanabe, and T. Takano, Applications of the
    Compliance Concept in Fracture Mechanics, in: "Progress in
    Flaw Growth and Fracture Toughness Testing," ASTM STP-536,
    American Society for Testing and Materials, Philadelphia
    (1973).
7. "Fracture Toughness Evaluation by R-Curve Methods," ASTM
    STP-527, American Society for Testing and Materials,
    Philadelphia (1973).
8. S. Gaggar and L. J. Broutman, Crack Growth Resistance of Random
    Fiber Composites, J. of Comp. Materials, 9:216 (1975).
9. W. F. Brown and J. E. Srawley, "Plane Strain Crack Toughness
    Testing of High Strength Metallic Materials," ASTM STP-410,
    American Society for Testing and Materials, Philadelphia
    (1966).
10. J. E. Srawley and W. F. Brown, Fracture Toughness Testing
    Methods, in: "Fracture Toughness Testing and Its Applica-
    tion," ASTM STP-381, American Society for Testing and
    Materials, Philadelphia (1965).
11. S. Mostovoy and E. J. Ripling, Fracture Toughness of An Epoxy
    System, J. of Applied Polymer Science, 10:1351 (1966).

12. T. R. Brusset and T. Chiu, "Fracture Mechanics for Structural Adhesive Bonds - Final Report," AFML-TR-77, Air Force Materials Laboratory, Wright-Patterson AFB (1977).

13. M. J. Salkind, Fatigue of Composite Materials, in: "Composite Materials: Testing and Design (Second Conference," ASTM STP-497, American Society for Testing and Materials, Philadelphia (1972).

14. J. J. Nevadunsky, J. J. Lucas, and M. J. Salkind, Early Fatigue Damage Detection in Composite Materials, J. of Comp. Materials, 9:394 (1975).

15. K. L. Reifsnider, W. W. Stinchcomb, and T. K. O'Brien, Frequency Effects on a Stiffness-Based Fatigue Failure Criterion in Flawed Composite Specimens, in: "Fatigue of Filamentary Composite Materials," K. L. Reifsnider and K. N. Lauraitis, eds., ASTM STP-636, American Society for Testing and Materials, Philadelphia (1977).

16. W. S. Johnson, Characterization of Fatigue Damage Mechanism in Continuous Fiber Reinforced Metal Matrix Composites, Ph.D. Dissertation, Duke University, Durham (1979).

17. W. W. Stinchcomb, K. L. Reifsnider, and R. S. Williams, Critical Factors for Frequency Dependent Fatigue Processes in Composite Materials, Experimental Mech., 16(9):343 (1976).

18. H. T. Hahn and R. Y. Kim, Fatigue Behavior of Composite Laminate, J. of Comp. Materials, 10:156 (1976).

19. T. K. O'Brien, An Evaluation of Stiffness Reduction as a Damage Parameter and Criterion for Fatigue Failure in Composite Materials, Ph.D. Dissertation, Virginia Polytechnic Institute and State University, Blacksburg (1978).

20. T. K. O'Brien and K. L. Reifsnider, A Complete Mechanical Property Characterization of a Single Composite Specimen, Experimental Mech., 20(5):145 (1980).

21. G. L. Roderick, Stiffness Change During Fatigue of Composite Materials, in: "Proceedings of the Fifth Annual Mechanics of Composites Review," AFWAL-TR-80-4020, Air Force Wright Aeronautical Laboratories, Wright-Patterson AFB (1980).

22. T. K. O'Brien, Delamination Growth in Composite Laminates, to be presented at the ASTM Symposium on Damage in Composite Materials: Basic Mechanisms, Accumulation, Tolerance, and Characterization, Bal Harbour (1980).

CONCEPTS AND TECHNIQUES FOR ULTRASONIC EVALUATION

OF MATERIAL MECHANICAL PROPERTIES

Alex Vary

National Aeronautics and Space Administration
Lewis Research Center
Cleveland, Ohio 44135

INTRODUCTION

Reliable performance of advanced, high-strength materials in
critical applications depends on assuring that each part placed in
service satisfies the conditions assumed in design and life prediction analyses. Reliability assurance requires the availability
of nondestructive evaluation (NDE) techniques not only for defect
detection but also for verification of mechanical strength and
associated properties. Advanced NDE techniques are needed to con-
firm that metallic, composite, or ceramic parts will not fail under
design loads due to inadequate or degraded mechanical strength.
This calls for NDE techniques that are sensitive to variations in
microstructure, extrinsic properties, and dispersed flaw populations
that govern the ultimate mechanical performance of a structure.

The purpose of this paper is to review ultrasonic methods that
can be used for material strength prediction and verification.
Emergent technology involving advanced ultrasonic techniques and
associated measurements is described. It is shown that ultrasonic
NDE is particularly useful in this area because it involves mechanical elastic waves that are strongly modulated by morphological
factors that govern mechanical strength and also dynamic failure
modes. These aspects of ultrasonic NDE will be described in conjunction with advanced approaches and theoretical concepts for
signal acquisition and analysis for materials characterization.
It is emphasized that the technology is in its infancy and that much
effort is still required before the techniques and concepts can be
transferred from laboratory to field conditions.

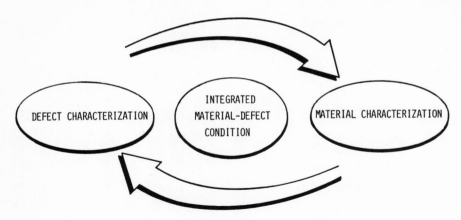

Fig. 1. Diagram illustrating the relation of defect and material
         characterization to defining the integrated effect of the
         material-defect state on structural integrity and life.

PRELIMINARY CONSIDERATIONS

     In its most general context, nondestructive evaluation (NDE) is
a branch of materials science that is concerned with all aspects of
the uniformity, quality, and serviceability of materials and structures. Therefore, NDE should not be defined solely by the current
emphasis on the detection of overt flaws (Sharpe, 1976). Certainly,
it is necessary to extend NDE technology to characterize discrete
flaws according to their location, size, orientation, and nature.
This leads to improved assessment of the potential criticality of
individual flaws. Concurrently, it is necessary to develop NDE
techniques for characterizing various inherent material properties.
In this case, the emphasis is on evaluation of microstructural and
morphological factors that ultimately govern mechanical strength and
dynamic performance. As illustrated in Fig. 1, a holistic approach
combines nondestructive characterization of defects and also material
environments in which the defects reside. This leads to improved
accuracy in predicting structural integrity and life upon exposure to
service conditions, particularly in the presence of discrete flaws.

     The specification of flaw criticality and prediction of safe
life depend on the assumption of a realistic set of extrinsic properties and conditions, such as those listed in Fig. 2. Fracture and
life prediction analysis models invariably presuppose flaw development and propagation in materials with well established moduli,
ultimate strengths, fracture toughnesses, and fatigue and creep properties. It is within the province and capability of NDE technology
to verify whether or not a structural part possesses the properties
assumed in design analysis (Vary, 1980). There are numerous NDE

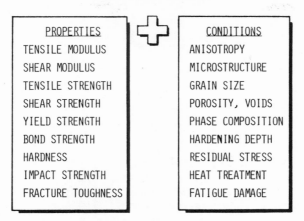

Fig. 2.   Material properties and conditions that can be assessed
by various nondestructive evaluation (NDE) techniques.

techniques that can be used for material properties characterization
(e.g., radiometric, electromagnetic, ultrasonic)(McMaster, 1959;
Green, 1973; Krautkramer, 1977; Hayward, 1978).   Many of these are
complementary and can be used to extend or corroborate measurements
by other methods.   This paper focuses on ultrasonic techniques that
have demonstrated potentials for materials characterization.   These
techniques rely on physical acoustic properties of materials and the
interaction of elastic stress waves with morphological factors in the
ultrasonic regime (Mason, 1958; Kolsky, 1963; Kolsky, 1973).

All the material properties and conditions listed in Fig. 2 are
amenable to ultrasonic evaluation to differing degrees (Vary, 1978a;
1980).   The speed of wave propagation and energy loss by interaction
with material microstructure and geometrical factors underlie ultrasonic determination of material properties.   There is a well-established body of theoretical and experimental knowledge concerning the
ultrasonic measurement of elastic moduli (Truell et al, 1969; Schreiber et al, 1973).   Conversely, ultrasonic prediction of tensile and
yield strengths, and fracture toughness are currently based on empirical correlations (Vary, 1978b).

Proposed models for explaining the above-mentioned empirical
correlations invoke the concept of ultrasonic stress wave interactions with material microstructure to the degree where the stress
waves actually promote plastic deformation and microcrack extension
(van Elst, 1973; Vary, 1979a).   This stress wave interaction concept
forms the basis for an ultrasonic approach to defining material--
defect interactions as a means for prediction of ultimate strength
and dynamic reaction to applied loads.   Illustrative examples of the
concept are discussed hereinafter.

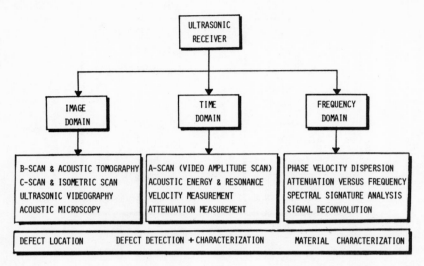

Fig. 3.   Alternative data processing and analysis methods upon the
acquisition of ultrasonic signals from a test article.

ULTRASONIC DOMAINS

There are three major domains for presenting, processing, and
analyzing ultrasonic data: (i) image domain, (ii) time domain, and
(iii) frequency domain. As indicated in Fig. 3, the detailed treatment of ultrasonic signals within each domain can be accomplished by
various methodologies, e.g., acoustic tomography, acoustic microscopy,
velocity and attenuation measurement, spectral signature analysis
(Brown, 1973; Kessler and Yuhas, 1978; Krautkramer, 1977; Vary 1980).
The end objectives range from defect detection to material property
characterization.

Irrespective of the methodology used, the fundamental process
in the image domain produces a representation of signal strength
against spatial coordinates. An example is given in Fig. 4 wherein
material quality variations associated with microvoids and fiber
content in a composite laminate are revealed. In the image domain,
the location and size of flaws or the extent of defective material
become apparent. The chief advantage of ultrasonic imaging is in
affording means for qualitative ranking of test articles relative to
defect populations and material anomalies (Posakony, 1978).

The time domain methodologies all employ electrical analogs of
ultrasonic echoes and transmitted waveforms that are displayed as
signal amplitude versus time oscilloscope traces. Specific signals
are selected for detailed examination and quantitative measurements
of energy, velocity, or attenuation. Time domain measurements are
currently predominant in defect and material characterization.

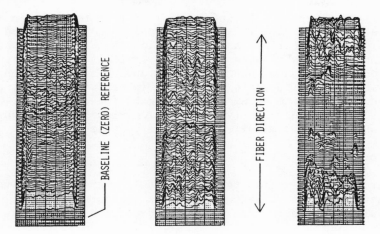

Fig. 4. Through transmission immersion ultrasonic amplitude scans
         (isometric scans) of graphite/polyimide composite laminate
         panels. Scans show variations of transmitted signal rel-
         ative to zero transmission baseline reference at bottom.
         Although each panel was formed with the same cure pressure,
         it is evident that material quality and uniformity differ
         from panel to panel (Vary and Bowles, 1979).

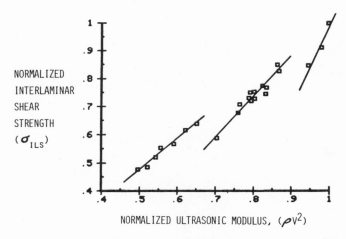

Fig. 5. Interlaminar shear strength of graphite/polyimide composite
         laminate specimens compared to ultrasonic modulus based on
         density and through-thickness velocity measurements. Three
         separate correlation curves that were obtained corresponded
         to different combinations of morphological factors that con-
         trolled fracture modes during short beam shear tests for
         interlaminar shear strength (Vary and Bowles, 1977).

Material strength correlations derived from time domain signals are indicated in Fig. 5. This is an example of a widely-used ultrasonic approach to material characterization that involves measuring elastic constants and related strength properties. Measurement of elastic moduli are fundamental to understanding and predicting material behavior. Since they are related to interatomic forces, elastic moduli indicate maximum attainable strengths (Green, 1973). Longitudinal (v_1) and transverse (v_t) wave velocities give the longitudinal (L) and shear (G) moduli, respectively, where,

$$L = \rho v_1^2 \quad \text{and} \quad G = \rho v_t^2 \tag{1}$$

For linear isotropic solids these two moduli are sufficient to completely define elastic behavior, given interconnecting relations with other moduli, e.g., bulk modulus, tensile modulus, Poisson's ratio and the Lamé constant (Schreiber et al, 1973). Anisotropic and most polycrystalline solids present a more complex situation since the principal moduli (L and G) will assume different values with different directions of ultrasonic wave propagation. Nevertheless, there exists an extensive literature that confirms the capabilities of various time domain measurements for predicting mechanical strength for materials ranging from cast iron to concrete (Vary, 1980).

Frequency domain methodologies begin with the acquisition and transformation of time domain signals. The transformations to the frequency domain are made by either (i) analog frequency spectrum analysis or (ii) digital Fourier transform algorithms (Gericke, 1970; Adler et al, 1977; Rose and Thomas, 1979; Vary, 1979b). Working in the frequency domain affords access to defect and material characterization data that are unattainable or impractical to seek in the time domain. An example of the frequency domain approach and methodology is discussed under MATERIAL TRANSFER FUNCTION.

STRESS WAVE INTERACTION

As mentioned previously, significant correlations of ultrasonic attenuation and velocity with material strength properties exist. Many of these ultrasonic versus property correlations appear to be fortuitous, having been found by trial or chance rather than by extentions of established principles. The classical elastic wave model does support the expectation that velocity will relate to strength through elastic moduli. However, current theory does not adequately account for the strong correlations of ultimate strengths and fracture toughness with attenuation. It is proposed that this lack can be remedied by considering fracture models in which ultrasonic stress waves interact with material morphological factors to the extent that they actually promote microcracking and also catastrophic crack extention. This point of view coincides with dynamically-based models for fracture behavior (Kolsky, 1973; Curran et al, 1977).

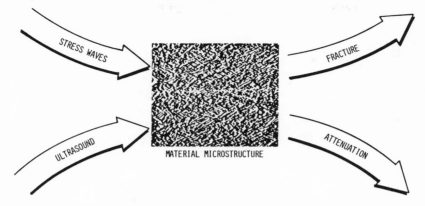

Fig. 6.   Depiction of the equivalence of ultrasound and stress
          wave propagation under linear elastic conditions wherein
          material microstructure governs ultrasonic attenuation
          and fracture phenomena.

     The stress wave interaction concept stated above can be used for
developing a theoretical basis for correlations found between ultrasonic attenuation and material strength and toughness. The working
hypothesis is that given linear elastic conditions, propagation of
probe ultrasound is governed by the same material morphological
factors that govern stress waves generated during fracture, Fig. 6.
The importance of microstructure in controlling mechanical behavior
is, of course, well established (MacCrone, 1977; Froes et al, 1978).
The use of probe ultrasound, as depicted in Fig. 7, would be expected
to define material transfer functions that determine stress wave interactions such as redirection and energy loss due to scattering and
absorption, for example. Considering material microstructure as a
filter with a transfer function definable in terms of the ultrasonic
attenuation coefficient proves to be a useful concept, as indicated
by results cited under FRACTURE TOUGHNESS AND ATTENUATION.

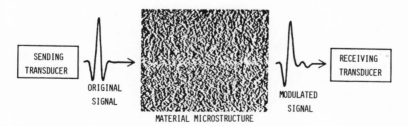

Fig. 7.   Depiction of material microstructure as an ultrasonic
          wave filter in which a standard reference signal becomes
          modulated according to a definable transfer function.

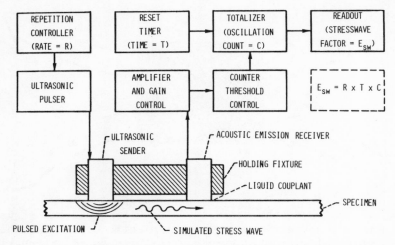

Fig. 8.   Diagram of acousto-ultrasonic apparatus for measurement of
          the stress wave factor $E_{sw} = R \cdot T \cdot C$.   The quantity C is the
          number of time domain "ringdown" oscillations exceeding a
          threshold voltage as in the acousto-ultrasonic waveform
          shown in Fig. 9 (Vary and Bowles, 1977; 1979).

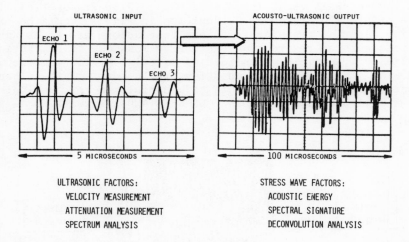

Fig. 9.   An ultrasonic pulsed input (left) is used to excite the
          acousto-ultrasonic output waveform (right) from which the
          stress wave factor, E_{sw}, is measured.   Both the ultrasonic
          input pulse echoes and acousto-ultrasonic output can be
          measured by the alternative factors indicated in order to
          determine the material modulation transfer function.

STRESS WAVE FACTOR

An illustrative example of the application of the stress wave interaction concept is given herewith. The application involves a novel approach that was developed to evaluate fiber composite panels for mechanical strength properties and in-service strength loss. The approach combines instrumentation from two previously separate technologies: (i) acoustic emission and (ii) pulse ultrasonics (Liptai and Harris, 1971; Spanner, 1974; Krautkramer, 1977). The usual procedure with acoustic emission involves the detection and analysis of spontaneous stress wave emissions due to material defor- mation and flaw growth. The "acousto-ultrasonic" procedure employs ultrasonically excited elastic waves that simulate acoustic emission events, as indicated in Figs. 8, 9 (Vary and Bowles, 1977; 1979).

The object is to generate a repeating, controlled set of elastic waves that will interact with material morphology and boundary sur- faces in a manner similar to spontaneous stress waves that arise at the onset of fracture. The resultant output waveform resembles "burst" type acoustic emission both in the time and frequency domains. Like spontaneous acoustic emission waveforms the acousto-ultrasonic waveform carries substantially more information on the material in which it runs than on the signal source. It is a mixed function of multimode velocities, attenuations, dispersions, and reflections. It has been demonstrated that, in the restricted case of fiber composite laminates, the acousto-ultrasonic waveform will yield correlations with ultimate tensile and interlaminar shear strengths, Figs. 10, 11.

The correlations were obtained by measurement of a "stress wave factor" (see Fig. 8). The stress wave factor may be described as a measure of the efficiency of stress wave energy transmission. This factor apparently provides a means for rating the efficiency of the dynamic strain energy transfer in the composites tested heretofore (Vary and Lark, 1979). Once microcracking starts in the brittle matrix or fibers, it is to be expected that prompt dissipation of stress wave energy away from the crack initiation sites contributes to dynamic integrity and ultimate strength. In unidirectional com- posites, the stress wave factor is greatest along the fiber direction which is also the direction of maximum strength. Regions of small values of stress wave factor are regions of higher ultrasonic atten- uation (Williams and Lampert, 1980). These regions are also observed to be regions of weakness where dynamic strain energy is likely to concentrate and promote further microcracking failure.

The preceding discussion leads to a point made previously with regard to the phenomenon of stress wave interactions and their rela- tion to failure dynamics. The fundamental argument being advanced is that spontaneous stress waves that arise during microcracking can interact with other potential crack sites leading to either cleavage

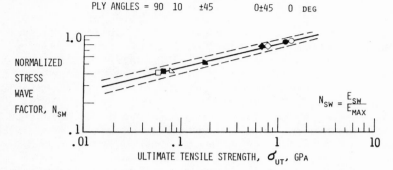

Fig. 10. Stress wave factor as a function of ultimate tensile
          strength for graphite/epoxy fiber composite. The stress
          wave factor is normalized relative to its maximum value
          for the specimen materials. The ply angles given are
          relative to the loading axis (Vary and Lark, 1979).

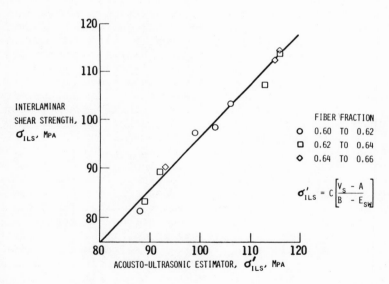

Fig. 11. Correlation of graphite/polyimide fiber composite inter-
          laminar shear strength with acousto-ultrasonic estimator.
          The estimator is derived from stress wave factor and vel-
          ocity measurements, E_{sw} and V_s, respectively. A, B, and
          C are experimental constants (Vary and Bowles, 1977).

or void coalescence and thence large-scale abrupt failure, provided an initiating excess strain has been applied (Vary, 1979a). Since the stress waves are ultrasonic in nature and subject to modulation by the material microstructure, it should be possible to determine a modulation transfer function by ultrasonic means. Measurement of a stress wave factor as described herein affords only a relative means. Time domain attenuation measurements provide alternative means if the material sample geometry permits access along appropriate directions. However, the more appropriate approach is to work in the frequency domain wherein signal deconvolution is readily accomplished and the material transfer function can be precisely defined.

MATERIAL TRANSFER FUNCTION

The conditions under which the material transfer function can be defined are restricted. An isotropic polycrystalline aggregate is assumed for the purposes of this discussion. It is also assumed that the sample has flat, parallel opposing surfaces and satisfies the conditions necessary to obtain two back surface echoes as indicated in Fig. 12 (Truell et al, 1969). Signal acquisition and processing would be accomplished as indicated in Fig. 13 (Vary, 1979b).

It will be seen that frequency domain analysis yields an ultrasonic transfer function, T, for the material in terms of its attenuation coefficient, α, and reflection coefficient, R. The quantities B1, B2, E1, E2, T, and R are taken as Fourier transforms of corresponding time domain quantities (Bracewell, 1978). This puts the aforementioned quantities into the frequency domain where signal deconvolution and transfer function definition can proceed with simple mathematical manipulations. The attenuation coefficient, being a function of frequency, is likewise defined in the frequency domain,

$$\alpha = cf^m \qquad\qquad (2)$$

where, f is frequency and c and m are experimental constants (Vary, 1978b; Serabian, 1980), given that scatter attenuation prevails.

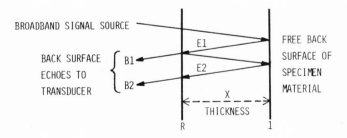

Fig. 12.   Diagram of echo system showing quantities involved in the definition of the material ultrasonic transfer function.

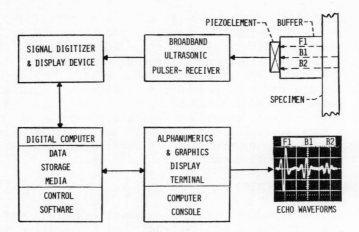

Fig. 13. Diagram of system for ultrasonic signal acquisition and
          analysis in time and frequency domains (Vary, 1979b).

As indicated in Fig. 12, a broadband ultrasonic pulse signal produces a series of back surface echoes in the material specimen. The
first two back surface echoes B1 and B2 re-enter the ultrasonic transducer which acts as sender and receiver, Fig. 13. It is appropriate
to take the internal echo E1 as the source signal for B1, thus,

$$B1 = (1+R)E1 \qquad\qquad (3)$$

where, (1+R) is the transmission function at the specimen-transducer
interface (Truell, et al, 1969). A portion of the energy of E1 is reflected and appears as the second internal echo E2, giving,

$$B2 = TR(1+R)E1 \qquad\qquad (4)$$

where, the transfer function T incorporates signal modulation factors
associated with the material microstructure (e.g., grain scattering,
absorption, etc.) and interface effects. Combining the two preceding
equations,

$$T = B2/RB1 \qquad\qquad (5)$$

The transfer functions associated with coupling and other factors of
signal transduction were ignored as they cancel out just as the term
(1+R)(E1) vanishes upon combining Equations (3) and (4) to get (5).
It has been shown by Papadakis (1976) that the attenuation coefficient
can be measured by frequency spectrum analysis and that,

$$\alpha = (1/2x) \; \ln(RB1/B2) \qquad\qquad (6)$$

where, x is specimen thickness.

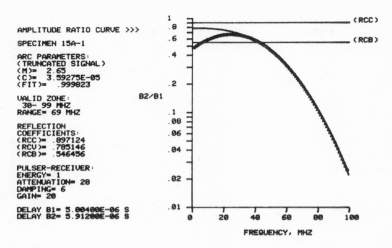

Fig. 14.   Computer documentation of amplitude ratio curve and data
           associated with signal acquisition and analysis. The ratio
           B2/B1 as a function of frequency is the ultrasonic transfer
           function of the specimen material (Vary, 1979b).

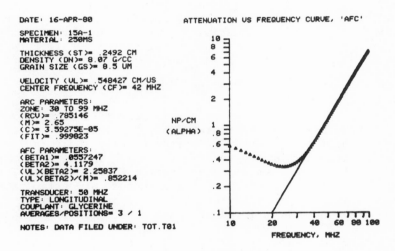

Fig. 15.   Computer documentation of attenuation as a function of
           ultrasonic frequency and associated data. The straight
           line (through the raw data derived from Fig. 14) is the
           computed attenuation vs. frequency curve for the specimen
           material (Vary, 1979b).

By noting that the ratios, B2/B1, appearing in Equations (5) and (6) are identical functions of ultrasonic frequency, we have,

$$T = \exp(-2x\alpha) \tag{7}$$

that is, the material transfer function or ultrasonic wave filtering characteristic is defined in terms of the attenuation coefficient and reflection coefficient. For nondispersive materials, the reflection coefficient R is independent of frequency. It is a function of material velocity and density (Truell et al, 1969; Papadakis, 1976). Recalling that B1 and B2 were taken as Fourier transforms of corresponding time domain echoes, it is clear that Equation (5) gives the transfer function T as the ratio of the frequency spectra of time domain waveforms. Therefore, in complex polar form,

$$T = (1/R)(a2/a1)\exp(i\phi) \tag{8}$$

where, a1 and a2 are the amplitude spectra for signals B1 and B2, respectively, while ϕ is the difference in phase spectra ($\phi2-\phi1$). Here, T represents the deconvolution of the time domain counterparts of B1 and B2 (Newhouse and Fugason, 1977; Bracewell, 1978).

Equations (7) and (8) are a basis for determining material properties by means of ultrasonic spectrum analysis and associated ultrasonic attenuation measurement. The essential operations for accomplishing this, as implemented by a computer system, are illustrated in Figs. 14 and 15. A number of ultrasonic factors derived from material transfer function and attenuation curves have proven to correlate well with microstructure, fracture toughness, and yield strength in metals, as discussed in the following section.

FRACTURE TOUGHNESS AND ATTENUATION

The feasibility of ultrasonic measurement of plane strain fracture toughness has been demonstrated for two maraging steels and a titanium alloy (Vary, 1978b). A principal ultrasonic factor that correlates with fracture toughness is β which is the slope, $d\alpha/df$, of the attenuation versus frequency curve, Equation (2). The constants c and m for the material microstructure are established by the frequency domain analyses represented in Figs. 14 and 15. The correlations that have been found are shown in Figs. 16, 17, and 18.

Fracture toughness, yield strength, and related microstructural factors are apparently intimately connected with ultrasonic and hence (stress) wave propagation factors in polycrystalline metallic materials. The empirical correlations that are exhibited in Figs. 16 through 18 imply that stress wave interactions are important during rapid (catastropic) crack extension, as under the conditions for determining fracture toughness (Brown and Srawley, 1966).

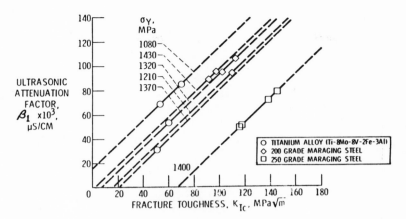

Fig. 16.   Ultrasonic attenuation factor β_1 as a function of fracture toughness K_{Ic} and yield strength σ_Y for three metals.   The material specimens that share the same yield strength are represented on the same line, for example, the 250-grade maraging steels with $\sigma=1400$ MPa (Vary, 1978b).

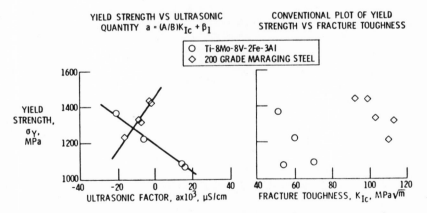

Fig. 17.   Correlations of yield strength to fracture toughness for a titanium alloy and a maraging steel.   The lefthand graph is based on data from Fig. 16 and combines the ultrasonic factor β_1 with fracture toughness K_{Ic} in the quantity a as defined above the figure (Vary, 1978b).

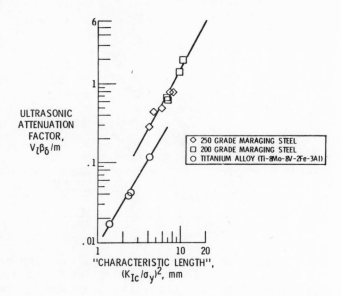

Fig. 18. Experimental correlation of ultrasonic attenuation factor
and fracture toughness "characteristic length" factor for
three metals. The experimental data agree with theoretical
relation given by Equation (9) (Vary, 1979a).

It can be inferred that spontaneous stress waves generated during crack nucleation will contribute to promoting the onset of rapid
unstable crack extension. A stress wave interaction model based on
this idea was used to derive equations that predict the empirical
correlations shown in Figs. 17 and 18 (Vary, 1979a). For the polycrystalline aggregates for which the equations were derived there exists a close relation between fracture toughness and yield strength.
This accounts for the appearance of yield strength σ_y in the equations
connecting plane strain fracture toughness K_{Ic} and the ultrasonic
attenuation factor β,

$$(K_{Ic}/\sigma_y)^2 = M(v_\ell \beta_\delta/m)^{\frac{1}{2}} \tag{9}$$

$$\sigma_y \mp AK_{Ic} \pm B\beta_1 = C \tag{10}$$

where, v_ℓ is velocity and A, B, C, and M are experimental constants
that are related to material microstructural factors. The quantity β_1
is the derivative $d\alpha/df$ evaluated at an attenuation coefficient $\alpha=1$,
while β_δ is $d\alpha/df$ evaluated at a particular threshold frequency that
corresponds to a critical ultrasonic wavelength in the material. This
wavelength is related to the mean grain boundary spacing. Equation
(9) describes the lines through the data in Fig. 18 while equation

(10) describes the lines through the data in Fig. 17.   The empirical coefficient A and B in Equation (10) carry opposite algebraic signs that appear to depend on the mode of fracture.   Thus, if these co-efficients are experimentally determined for a material that frac-tures in a predominantly brittle manner, A assumes a negative sign, giving a negative slope as for the line for titanium in Fig. 17. The coefficients and associated quantities in Equation (10) apparent-ly relate to modes of stress wave energy dissipation, residual strain in crack nucleation sites, and whether the nucleation sites are energy "sinks" or "sources" during fracture.   The coefficient M in Equation (9) appears to be related to microstructural factors such as grain size, lath spacing, ligament length (Hahn et al, 1972).   The quantity m in Equation (9) is the exponent on frequency in Equation (2).   Once these experimental constants have been determined for a material, Equations (9) and (10) can be taken as simultaneous rela-tions to solve for K_{Ic} and σ_y in terms of the ultrasonic factors.

CONCLUDING REMARKS

     The ultrasonic NDE approaches and results that have been high-lighted herein indicate potentials for material characterization and property prediction.   Stress wave interaction and material transfer function concepts were cited as bases for explaining correlations be-tween material mechanical behavior and ultrasonically-measured quanti-ties.   It is observed that the criticality and effect of any discrete flaw (crack, inclusion, or other stress raiser) is definable only in terms of its material microstructural environment.   This underscores the importance of ultrasonic techniques that can characterize stress wave energy transfer properties of a material.

REFERENCES

Adler, L., Cook, K. V., and Simpson, W. A., 1977, Ultrasonic frequen-
     cy analysis, in: "Research Techniques in Nondestructive Test-
     ing," Vol. 3, R. S. Sharpe, ed., Academic Press, London.
Bracewell, R. N., 1978, "The Fourier Transform and Its Applications,"
     McGraw-Hill, New York.
Brown, A. F., 1973, Materials testing by ultrasonic spectroscopy,
     Ultrasonics, 11:202.
Brown, Jr., W. F., and Srawley, J. E., 1966, "Plane Strain Crack Tough-
     ness Testing of High Strength Metallic Materials," ASTM STP 410
     Amer. Soc. for Testing and Materials, Philadelphia.
Curran, D. R., Seaman, L., and Shockey, D. A., 1977, Dynamic failure
     in solids, Physics Today, 30, 1:202.
Froes, F. H., Chesnutt, J. C., Rhodes, C. G., and Williams, J. C., 1978,
     Relationship of fracture toughness and ductility to microstruc-
     ture, in: "Toughness and Fracture Behavior of Titanium," R. G.
     Broadwell and C. F. Hickey, Jr., eds., ASTM STP 651, Amer. Soc.
     for Testing and Materials, Philadelphia.

Gericke, O. R., 1970, Ultrasonic spectroscopy, in: "Research Tech-
    niques in Nondestructive Testing," R. S. Sharpe, ed., Academic
    Press, London.
Green, Jr., R. E., 1973, Ultrasonic investigation of mechanical prop-
    erties, in: "Treatise on Materials Science and Technology,"
    Vol. 3, H. Herman, ed., Academic Press, New York.
Hahn, G. T., Kanninen, M. F., and Rosenfeld, A. R., 1972, Fracture
    toughness of materials, in: "Annual Reviews of Materials Sci-
    ence," Vol. 2, R. A. Huggins, ed., Annual Reviews, Inc.,
    Palo Alto.
Hayward, G. P., 1978, "Introduction to Nondestructive Testing," Amer.
    Soc. for Quality Control, Milwaukee.
Kessler, L. W. and Yuhas, D. E., 1978, Listen to structural differ-
    ences, Industrial Research/Development, 4:101.
Kolsky, H., 1963, "Stress Waves in Solids," Dover Publ., New York.
Kolsky, H., 1973, Recent work on the relation between stress pulses
    and fracture, in: "Dynamic Crack Propagation - Proceedings of
    an International Conference," G. C. Sih, ed., Noordhoff In-
    ternational Publ., Leyden.
Krautkramer, J., and Krautkramer, H., 1977, "Ultrasonic Testing of
    Materials," Springer-Verlag, Berlin.
Liptai, R. G., and Harris, D. O., 1971, "Acoustic Emission," ASTM STP
    505, Amer. Soc. for Testing and Materials, Philadelphia.
MacCrone, R. K., ed., 1977, Properties and microstructure, in: "Trea-
    tise on Materials Science and Technology," Vol. 11, Academic
    Press, New York.
McMaster, R. C., ed., 1959, "Nondestructive Testing Handbook," Ronald
    Press Co., New York.
Mason, W. P., 1958, "Physical Acoustics and the Properties of Solids,"
    D. Van Nostrand Co., Princeton.
Newhouse, V. L., and Furgason, E. S., 1977, Ultrasonic correlation
    techniques, in: "Research Techniques in Nondestructive Testing,"
    R. S. Sharpe, ed., Academic Press, London.
Papadakis, E. P., 1976, Ultrasonic velocity and attenuation measure-
    ment methods with scientific and industrial applications, in:
    "Physical Acoustics - Principles and Methods," Vol. 12, W. P.
    Mason and R. N. Thurston, eds., Academic Press, New York.
Posakony, G. J., 1978, Acoustic imaging - a review of current tech-
    niques for utilizing ultrasonic linear arrays for producing
    images of flaws in solids, in: "Elastic Waves and Nondestruc-
    tive Testing of Materials," Y. H. Pao, ed., Amer. Soc. of
    Mech. Engineers, New York.
Rose, J. L., and Thomas, G. H., 1979, The Fisher linear discriminant
    function for adhesive bond strength prediction, British J. of
    Non-Destructive Testing, 21, 3:135.
Schreiber, E., Anderson, O. L., and Soga, N., 1973, "Elastic Con-
    stants and Their Measurement, McGraw-Hill, New York.
Serabian, S., 1980, Frequency and grain size dependency of ultrasonic
    attenuation in polycrystalline materials, British J. of Non-
    Destructive Testing, 22, 2:69.

Sharpe, R. S., 1976, Innovation and opportunity in NDT, British J. of Non-Destructive Testing, 18, 4:98.

Spanner, J. C., 1974, "Acoustic Emission - Techniques and Applications," Intex Publ., Co., Evanston.

Truell, R., Elbaum, C., and Chick, B. B., 1969, "Ultrasonic Methods in Solid State Physics," Academic Press, New York.

Van Elst, H. C., 1973, The relation between increase in crack arrest temperature and decrease of stress wave attenuation, in: "Dynamic Crack Propagation - Proceedings of an International Symposium," G. C. Sih, ed., Noordhoff International Publ., Leyden.

Vary, A., 1978a, Quantitative ultrasonic evaluation of mechanical properties of engineering materials, TM-78905, National Aeronautics and Space Administration, Washington, D.C.

Vary, A., 1978b, Correlations among ultrasonic propagation factors and fracture toughness properties of metallic materials, Materials Evaluation, 36, 7:55.

Vary, A., 1979a, Correlation between ultrasonic and fracture toughness factors in metallic materials, in: "Fracture Mechanics," C. W. Smith, ed., ASTM STP 677, Amer. Soc. for Testing and Materials, Philadelphia.

Vary, A., 1979b, Computer signal processing for ultrasonic attenuation and velocity measurements for material property characterizations, in: "Proceedings of the Twelfth Symposium on Nondestructive Evaluation," Amer. Soc. for Nondestructive Testing, Columbus, and Southwest Research Institute, San Antonio.

Vary, A., 1980, Ultrasonic measurement of material properties, in: "Research Techniques in Nondestructive Testing," Vol. 4, R. S. Sharpe, ed., Academic Press, London.

Vary, A., and Bowles, K. J., 1977, Ultrasonic evaluation of the strength of unidirectional graphite/polyimide composites, in: "Proceedings of the Eleventh Symposium on Nondestructive Evaluation," Amer. Soc. for Nondestructive Testing, Columbus, and Southwest Research Institue, San Antonio.

Vary, A., and Bowles, K. J., 1979, An ultrasonic-acoustic technique for nondestructive evaluation of fiber composite quality, Polymer Engineering and Science, 19, 5:373.

Vary, A., and Lark, R. F., 1979, Correlation of fiber composite tensile strength with the ultrasonic stress wave factor, J. of Testing and Evaluation, 7, 4:185.

Williams, Jr., J. H., and Lampert, N. R., 1980, Ultrasonic nondestructive evaluation of impact-damaged graphite fiber composite, CR-3293, National Aeronautics and Space Administration, Washington, D.C.

NEUTRON DIFFRACTION AND SMALL-ANGLE SCATTERING AS NONDESTRUCTIVE PROBES OF THE MICROSTRUCTURE OF MATERIALS

C. J. Glinka,* H. J. Prask+* and C. S. Choi+*

*National Measurement Laboratory
 National Bureau of Standards
 Washington, D.C. 20234
+Energetic Materials Division
 LCWSL, ARRADCOM
 Dover, NJ 07801

INTRODUCTION

Thermal neutron scattering is a well established research tool for determining the atomic structure and fundamental excitations of solids and liquids. In addition, wide-angle diffraction has been employed for many years for the study of appropriate technological problems. In recent years, a growing area of interest in this field has been the development of scattering techniques which extend the scale of the structural information obtained to dimensions which are typical of the microstructure of a material as compared to its atomic structure. Such techniques, of which small-angle neutron scattering (SANS) is a prime example, yield information which is complementary with that obtained by electron microscopy or analogous x-ray methods. The neutron techniques afford the added advantage that the measurements may be made on relatively thick specimens or on samples in specialized environments owing to the highly penetrating nature of neutron radiation. The derived information is thus characteristic of the bulk material and can be correlated with bulk measurements carried out, in many cases, on the very same samples. Neutron scattering, therefore, has a natural role to play in the nondestructive investigation of materials properties.

In this article we first give a brief introduction to the fundamentals of neutron scattering, emphasizing those aspects which differ from analogous x-ray techniques. We then discuss small-angle neutron scattering methodology and applications pertinent to the nondestructive evaluation of materials properties. This is followed by a discussion of wide-angle diffraction techniques developed at NBS for the study of texture and residual stress in metallurgical samples.

SCATTERING FUNDAMENTALS

In any diffraction experiment, whether large-angle or small-angle diffraction, the measured scattered intensity is proportional to the square of a function $F(\vec{Q})$, usually referred to as the structure factor. This function can be expressed quite generally as

$$F(\vec{Q}) = \int_V \rho(\vec{r}) \, e^{i \vec{Q} \cdot \vec{r}} \, d\vec{r} \qquad (1)$$

where the integration is over the entire volume of the scattering sample. Thus, what is directly measured in a diffraction experiment is the square of a Fourier transform of a "scattering density", $\rho(\vec{r})$, which contains all of the structural information for the sample. The Fourier transform variable $\vec{Q}$ is called the scattering vector and is simply related to the incident and scattered beam directions,

$$\vec{Q} = \vec{k}_i - \vec{k}_f$$

$$|\vec{Q}| = \frac{4\pi}{\lambda} \sin(\theta/2) \qquad (2)$$

where the scattering angle θ is the angle between the incident and scattered wave vectors $\vec{k}_i$ and $\vec{k}_f$ ($k = 2\pi/\lambda$). The fundamental problem in diffraction is to infer $\rho(\vec{r})$ through a careful measurement of $|F(\vec{Q})|^2$.

In the case of x-rays, the density $\rho(\vec{r})$ is just the electron density of the specimen. Neutrons, however, scatter from the nuclei of atoms and the appropriate $\rho(\vec{r})$ in this case is called the scattering length or scattering amplitude density which may be written as $\rho(\vec{r}) = \sum_i \bar{b}_i \, \delta(\vec{r} - \vec{r}_i)$ where $\bar{b}_i$ is the coherent scattering amplitude of the ith atom located at $\vec{r}_i$. The concept of a coherent scattering amplitude is of such importance in neutron scattering that further explanation of this concept is warranted before continuing our discussion of diffraction.

Owing to the extremely short range of the neutron-nuclear interaction, a thermal neutron sees a nucleus as a point scatterer. As a result, the probability of a neutron being scattered by a nucleus is the same for all directions. This is expressed by

writing the scattered neutron wave, Ψ_s, assuming the incident
neutron is represented as a wave of amplitude 1.0 and wavelength λ,
as[1]

$$\Psi_s = -\left(\frac{b}{r}\right) \exp(2\pi ir/\lambda) \tag{3}$$

where r is the distance from the scattering nucleus to the observer.
Equation (3) serves to define the scattering length b. Its magnitude is a measure of the probability of the scattering event. The
scattered wave has a negative sign because it is usually 180° out
of phase relative to the incident wave.

The scattering length of an atom depends on its nuclear mass
and spin. Atoms with nuclear spin, in fact, have two scattering
lengths, often quite different: one for an encounter with a neutron
whose spin is up and another for a spin down encounter. Furthermore,
the scattering lengths of different isotopes of the same element
differ. Thus when a sample of an element scatters neutrons, part
of the scattering will reflect random variations in sample scattering
length due to the random distribution of nuclear spin orientations
and isotopic species. This random component is called incoherent
scattering, is spatially isotropic, and contains no structural
information about the sample. There will also be a component of
the scattering whose spatial distribution reflects the spatial
arrangement of atoms in the sample. This correlated component is
called coherent scattering. The cross section for coherent scattering per atom of an element is $\sigma_c = 4\pi(\bar{b})^2$ where the scattering
length averaged over the nuclear spin distribution and isotopic
abundance, $\bar{b}$, is what is meant by the coherent scattering amplitude.

The scattering length for x-rays, $f_x(\theta)$, which is equivalent
to $\bar{b}$ for neutrons, is defined from Equation (1) by carrying out the
integration only over the volume of a single atom. This scattering
length is not a constant, but decreases monotonically with increasing
scattering angle θ. More importantly, f_x is proportional to the
atomic number Z whereas $\bar{b}$ varies irregularly with Z. This behavior
is shown in Figure 1 in which values of $\bar{b}$ and f_x are plotted versus
Z. The point to be noted from Figure 1 is that light and heavy
elements have comparable scattering strengths for neutrons whereas
the heavy elements strongly dominate over light elements in scattering x-rays. As a result, determining the positions of light atoms
in heavy element hosts is often facilitated, and in some cases only
possible, with neutrons. Furthermore, because b (rather than $\bar{b}$)
varies irregularly with A (the atomic mass) as well as Z, isotopic
substitutions can often be used to aid in distinguishing the
contributions of individual elements in a neutron diffraction
pattern.

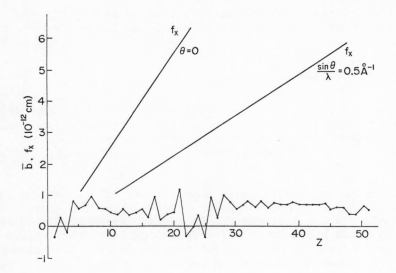

Fig. 1. Variation of neutron and x-ray scattering lengths with
atomic number Z. The elements with $\bar{b}<0$ are H, Li, Ti, V
and Mn.

Rather than being scattered, a neutron may be absorbed by a
nucleus with which it interacts to form a new nuclide. The neutron
absorption cross sections for most elements are small, however,
compared to the absorption of x-rays due to electronic excitations.
Typical absorption coefficients for neutrons[1] and x-rays are of
order 0.3 and 1000 cm^{-1}, respectively. As a result neutrons are,
in general, more than a thousand times more penetrating than x-rays.
This is the principal advantage that neutrons have with respect to
x-rays in the study of the bulk structural properties of materials.

Having identified the scattering density $\rho(\vec{r})$, we now return
to discuss its determination in a diffraction experiment. Generally
speaking, the intensity $I(Q)$ that one measures at a particular
scattering angle θ will be dominated by the Fourier components of
$\rho(\vec{r})$ with periodicities d such that $Qd \simeq 2\pi$ so that the phase
factor in Equation (1) is nearly unity. From Equation (2) this
condition becomes

$$2d \sin(\theta/2) \simeq \lambda \tag{4}$$

relating d to the scattering angle and the wavelength of the probing
radiation. Thus, in wide-angle scattering the shape of I(Q) reflects
structure in the sample on a scale d ~ λ (although the broadening
of peaks can be related to particle sizes). The determination of
crystal structures depends on the comparability of neutron and
x-ray wavelengths with the atomic spacings in crystalline materials.
In this case Equation (4) becomes a strict equality, Bragg's law,
and sharp diffraction peaks are observed which correspond to a
discrete set of well defined periodicities d, namely the spacings
between lattice planes.

Equation (4) applies equally well to small-angle scattering as
to the more familiar phenomena of wide-angle (Bragg) diffraction.[2]
The only difference is in the scale of the structural information
contained in the two regions of the scattering curve. The scattering
measured at very small angles arises from variations in $\rho(\vec{r})$ on
a scale which is large compared to the wavelength, i.e., d >> λ.
For example, at a scattering vector Q_o = 0.002 $Å^{-1}$, which corresponds
to a scattering angle of 0.13° for 7 Å neutrons, structure on the
order of d ≃ 2π/Q = 3000 Å can be studied. Few materials have well
defined periodicities on such a large scale, and as a result small-
angle scattering is not usually localized in Bragg peaks but is
diffuse in nature. Although the interpretation of such scattering
is not so direct and unambiguous as for Bragg scattering, considerable structural information, on a scale of vital importance to
understanding materials response, may often be derived from its
intensity and shape.

The structural range which can be explored by small-angle
scattering is limited by the spectrum of available wavelengths for
both neutrons and x-rays. Research reactors provide usable fluxes
of neutrons with wavelengths up to about 15 Å while the strong
absorption of x-rays places a practical limit of roughly 3 Å on
the usable range of wavelengths obtained from conventional x-ray
sources. Thus for identical beam collimation, somewhat smaller Q's
(larger d's) can be studied using neutrons. The longer neutron
wavelengths can provide a further advantage when studying crystalline
materials. By using wavelengths λ > $2d_m$, where d_m is the largest
lattice plane spacing, Bragg scattering is made impossible and
therefore multiple Bragg scattering, which peaks in the forward
direction, is eliminated. Multiple Bragg scattering may seriously
complicate x-ray small-angle scattering measurements.

Although used primarily for the study of crystal structure,
wide-angle diffraction techniques have been developed which are
sensitive to other aspects of structural properties. For example,
in a polycrystalline material the intensity of the Bragg peaks
depends on the number of crystallites which are oriented to satisfy
the Bragg condition [Equation (4)]. If there are preferred orientations of the crystallites (texture), a systematic study of the

intensities of various Bragg peaks as a function of sample orientation can reveal the macroscopic texture patterns in the sample.
Examples of texture studies, as well as other results related to
nondestructive testing, using wide-angle neutron diffraction are
discussed later in this article.

SMALL-ANGLE SCATTERING

Experimental Aspects

     The recent growth in the area of small-angle neutron scattering
(SANS) has hinged on the development of specialized instrumentation[3,4] which compensates for the relatively low intensity of
neutron sources [in terms of particles (or quanta) per unit time,
unit area, steradian, and wavelength interval] compared to x-ray
sources. The new generation of SANS instruments exploits the
diffuse nature of reactor sources by increasing the cross section
of the source aperture and the sample size while also elongating
the source-to-sample and sample-to-detector distances to maintain a
desired resolution. As a result the neutron instruments are massive
compared to their x-ray counterparts; the D11 instrument at the
Grenoble high-flux reactor has a maximum distance between entrance
slit and detector of 80 m.

     A second compensating feature of the neutron instruments is
the use of a relatively broad wavelength spread, $\Delta\lambda$, which can be
tolerated in small-angle scattering. The optimization requirement
that all contributions to the total resolution width of the scattering vector, δQ, have equal magnitudes dictates a wavelength spread
$\Delta\lambda/\lambda \simeq 0.15$, which is about 200 times larger than for x-ray tubes.

     Finally, the low attenuation of neutrons in materials already
discussed is a third factor which compensates for weaker source
strengths. Taken together these factors combine to make the best
of the neutron instruments fully competitive with current x-ray
techniques.

     As an example of the instrumentation used in SANS, we show in
Figure 2 a schematic layout of the new SANS spectrometer which is
presently under construction at the NBS reactor. This facility
utilizes a direct beam from the reactor which passes through two
liquid-nitrogen cooled filters consisting of 15 cm of single-
crystal bismuth to attenuate gamma rays and 25 cm of polycrystalline
beryllium which Bragg scatters out of the beam all wavelengths
$\lambda < 2d_m = 3.94$ Å, where d_m is the largest plane spacing for beryllium. The filtered beam then passes through a 40 cm long rotating
drum with helical channels cut along its length. This velocity
selector passes only a narrow band of neutron velocities

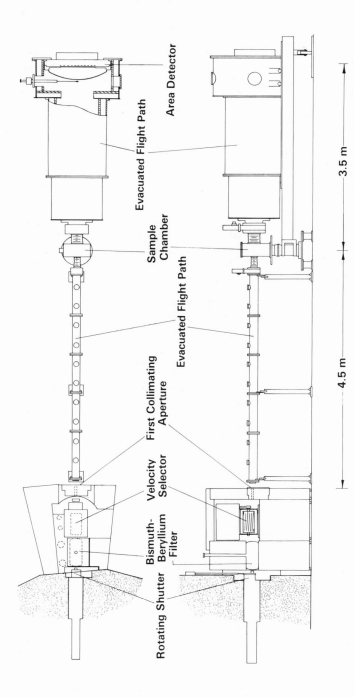

Fig. 2. A schematic layout of the new SANS instrument at NBS.

(wavelengths), whose mean value depends on the rate of rotation. At the NBS facility, two velocity selectors may be used interchangeably to vary the mean wavelength from 4 to 10 Å with a spread $\Delta\lambda/\lambda$ of either 0.05 or 0.20. Following the filters and velocity selector, the beam enters a 4.5 m evacuated flight tube with collimating apertures at either end to define the divergence of the incident beam, which may be varied from 5 to 15 minutes of arc. A second evacuated flight path after the sample houses the neutron detector which may be located at either 2.0 or 3.5 m from the sample. The detector and associated flight path mount on an arm which can rotate about the sample position to reach larger scattering angles.

A unique feature of the NBS instrument will be an optional high resolution collimation system, shown schematically in Figure 3, for use with larger samples (~1.5 cm x 1.5 cm). The collimation is done by a series of masks, having nine apertures each, which are aligned along the 4.5 m flight tube preceeding the sample. The effect of the masks is to produce nine independent beams which are geometrically focussed to a point at the center of the detector thereby increasing the beam area at the sample by nine while maintaining the resolution of a single beam channel. In this way a large area of the source is utilized without greatly increasing the size of the instrument.

Data over the entire small-angle region will be collected simultaneously on the NBS instrument with a large (65 x 65 cm^2) position-sensitive proportional counter of the Borkowski-Kopp type.[5] The resolution of this detector is 5 mm in both the horizontal and vertical directions giving a total of 128 x 128 detection elements. The 16,000 element data array is stored in the memory of

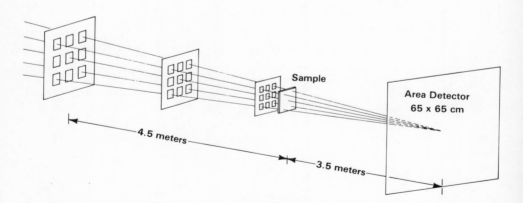

Fig. 3.   A representation of the nine channel, converging beam
          collimation system employed at the NBS SANS facility.

a dedicated minicomputer which is linked to a larger computer for data reduction and analysis as well as display on an interactive graphics terminal.

In a typical configuration with reasonably high resolution, the NBS SANS instrument will cover a Q range from 0.003 Å^{-1} to 0.10 Å^{-1} (with the beam centered on the detector) with a resolution ΔQ(FWHM) at the minimum Q of 0.002 Å^{-1}, assuming a wavelength of 6 Å and $\Delta\lambda/\lambda = 0.20$. At this resolution and wavelength, the flux on the sample is expected to be 6×10^4 n/cm^2-sec with the reactor operating at 20 MW. Other configurations of higher or lower resolution will allow fluxes on the sample between 10^4 to 10^6 n/cm^2-sec and access to wave vectors as small as 0.002 Å^{-1} and as high as 0.5 Å^{-1}.

The instrumentation and design parameters of the NBS SANS facility are comparable to those of other state-of-the art SANS instruments around the world including the NSF sponsored facility at Oak Ridge and the facilities at ILL in Grenoble, France.

SANS Applications

To date the bulk of the SANS studies relating to the nondestructive evaluation of materials properties has been carried out in Europe where the first modern SANS instruments were developed. For example, a considerable amount of work has been done by Pizzi, Walther and coworkers of the Fiat Corporation at the Galileo reactor in Pisa, Italy. Among other things this group has studied the γ' precipitates in nickel superalloys and in actual turbine blades made from these alloys. In order to discuss the results obtained, we must first reexamine Equation (1) as it applies to the case of scattering from widely dispersed particles in a homogeneous medium.

Consider for simplicity a single particle immersed in a uniform matrix.[2] Since the details of the atomic structure of the particle and matrix do not affect the small-angle scattering, the scattering length density $\rho(\vec{r})$ in Equation (1) may be replaced by its average value: $\rho_p(\vec{r}) = \bar{b}_p n_p$, $\rho_m(\vec{r}) = \bar{b}_m n_m$ where n_p and n_m are the number of nuclei per unit volume for the particle and matrix, respectively. Equation (1) then becomes

$$F(\vec{Q}) = \bar{b}_p n_p \int_{V_p} e^{i\vec{Q}\cdot\vec{r}} \, d\vec{r} + \bar{b}_m n_m \int_{V_m} e^{i\vec{Q}\cdot\vec{r}} \, d\vec{r} \qquad (5)$$

The second integral in Equation (5), over the volume of the matrix, is, however, just the negative of the first integral over the particle (Babinet's principle) so that

$$F(\vec{Q}) = (\bar{b}_p \, n_p - \bar{b}_m \, n_m) \int_{V_p} e^{i \, \vec{Q} \cdot \vec{r}} \, d\vec{r} \qquad (6)$$

involving only an integration over the volume of the particle.
Equation (6) may be applied to a real system containing many (widely
dispersed) particles of various sizes (but presumably similar
shapes) by averaging $|F(\vec{Q})|^2$ over all particle orientations and by
introducing a particle size distribution function N(R). The result
for the measured cross section is

$$\frac{d\Sigma}{d\Omega} = (\bar{b}_p \, n_p - \bar{b}_m \, n_m)^2 \int_0^\infty N(R) V^2(R) F_p^{\,2}(QR) \, dR \qquad (7)$$

where N(R) and V(R) are the number and volume of the particles with
average radius R and $F_p^{\,2}(QR)$ is the square of the integral in
Equation (6) averaged over all particle orientations.

   Although the exact form of the function $F_p^{\,2}$ in Equation (7)
depends on the shape of the particle, its asymptotic behavior at
both large and small values of Q is the same for all particle
shapes. For example, at small values of Q, such that QR << 1, $F_p^{\,2}$
has a Gaussian shape (Guinier's law)

$$F_p^{\,2} \propto \exp(-1/3 \, R_g^{\,2} \, Q^2) \qquad (8)$$

where R_g, the radius of gyration, is a measure of the overall size
of the particle.[2] For neutrons, R_g is defined by the expression,

$$R_g^{\,2} = \sum_i \bar{b}_i \, R_i^{\,2} / \sum_i \bar{b}_i \qquad (9)$$

similar to the definition of the radius of gyration in classical
mechanics, where $\vec{R}_i$ is the position of the ith atom in the particle
relative to an origin such that $\sum_i \bar{b}_i \, \vec{R}_i = 0$. At larger values of
Q, such that QR >> 1, $F_p^{\,2}$ has the simple behavior known as Porod's
law[2]

$$F_p^{\,2} \propto \frac{S}{Q^4} \qquad (10)$$

where S is the surface area of the particle. With general expressions such as these, it is often possible to extract essential
information about the scattering particles without having to analyze
the shape of the entire scattering curve.

The term preceding the integral in Equation (7) determines
the overall strength of the scattering and is referred to as the
contrast factor. It is this factor, and not simply $\bar{b}_p n_p$ alone,
which must be calculated when assessing the feasibility of an
experiment. Conversely, by measuring the scattering cross section
the contrast factor can be determined giving information on the
concentration of the constituent elements in the particle and
host.

An example of SANS measurements by the Fiat group[6] on heat
treated samples of Inconel X-750 as a function of aging time is
shown in Figure 4. Qualitatively, the gradual narrowing and
weakening of the scattered intensity indicates a coarsening of the
γ' particles with increasing aging time. These data were analyzed
in terms of Equation (7), assuming a spherical shape for the
particles, in order to determine the radius distribution function
N(R). To do this, the measured scattering curves were first
extrapolated to larger scattering angles using the general result

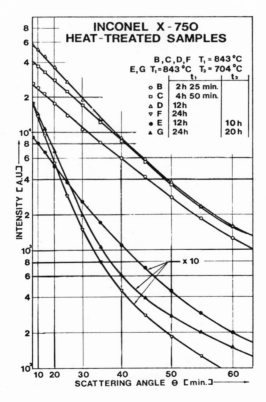

Fig. 4.   Small-angle scattering curves for heat-treated samples of
          Inconel X-750 (Ref. 6).

that F_p^2, the particle shape function, varies asymptotically as Q^{-4}, regardless of the particle shape (Porod's law). Equation (7) could then be inverted, using Mellon transforms, to obtain N(R) directly, without having to assume a parameterized form for its shape. The results of this procedure are shown in Figure 5. The curves in Figure 5 provide a quantitative description of the γ' coarsening process carried out nondestructively on a single specimen.

The results of similar SANS measurements by the Fiat group[6] on actual Inconel 700 turbine blades are shown in Figure 6. Summarized in the figure are the average radii of the γ' particles measured at four positions along the axes of the blades after various times of service. These data show that not only do the γ' particles coarsen with service, but that the coarsening rate is more rapid near the middle of the blades, particularly if the blade is operated above normal temperatures. Such results demonstrate the feasibility of using SANS to follow changes in the microstructure of actual mechanical components leading perhaps to the ability to correlate these changes with component performance.

Another area which has received considerable attention using SANS has been the study of voids in metals and alloys produced by irradiation with fast neutrons. For example, Frisius and Naraghi[7]

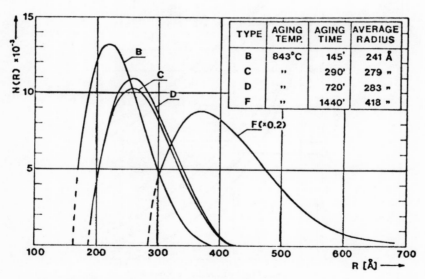

Fig. 5. The radii distribution functions for the precipitates in aged samples of Inconel X-750, derived from the data in Figure 4 (Ref. 6).

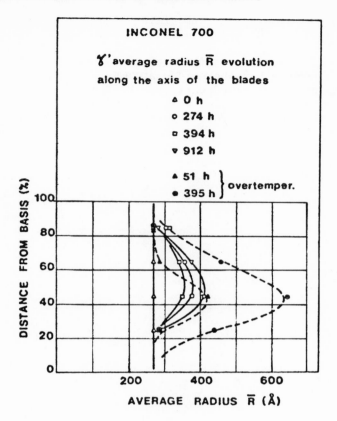

Fig. 6.   Evolution of the γ' average radius along the axis of
Inconel 700 turbine blades with time and temperature
(Ref. 6).

have examined the void scattering in reactor pressure vessel
steels.   In perhaps the most thorough work to date, Hendricks,
Schelten and coworkers[8] have studied the voids produced in single
crystals of aluminum and have compared the SANS results with
transmission electron microscope and immersion density measurements.
By making absolute measurements of the small-angle scattering, not
only the radii of gyration and void size distribution, but also
the swelling and total void surface area could be determined.   In
all cases good agreement was obtained with the other techniques.

Additional examples of recent materials science applications of
SANS may be found in the review articles in Ref. [9] and [10].

Some SANS measurements related to the nondestructive evaluation
of materials response have been made at NBS on a prototype to the
new SANS instrument described earlier. M. Fatemi[11] of the Naval
Research Laboratory has studied void formation associated with
creep in nichrome rods. He has observed that the void density

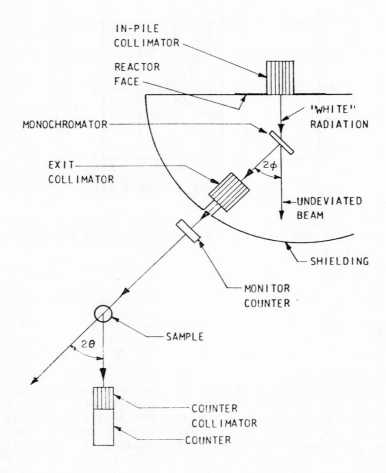

Fig. 7.   Schematic of typical wide-angle neutron diffractometer.

increases at low creep rates. Fatemi has also made preliminary
measurements on the growth and agglomeration of precipitates in
high strength low alloy steels as a function of heat treatment and
on the recrystallization process in a Ni-P glass. H. Alperin of
the Naval Surface Weapons Center has studied the strong small-angle
magnetic scattering which arises in a number of amorphous rare
earth-transition metal magnetic materials.[12] Because neutrons have
magnetic moments, they scatter from the magnetic moments of atoms
and thereby provide a unique and powerful tool for probing the
microscopic magnetic behavior of materials. By comparing data
taken in a magnetic field and in no field, Alperin has been able to
isolate the magnetic scattering and has shown that it originates
from unusually small magnetic domains (~80 Å). In addition, he has
demonstrated that these "microdomains" are a consequence of the
large single-ion magnetic anisotropy of the rare earth atoms and
not due solely to the amorphicity of the materials. SANS studies
in areas such as these are expected to be greatly expanded when the
new NBS facility comes into use.

WIDE-ANGLE NEUTRON DIFFRACTION

     As mentioned in the previous sections, two specific advantages
that neutron diffraction offers with respect to x-ray diffraction
for NDT studies are greater penetration and different scattering
selectivity. In the wide-angle diffraction range these advantages
can be utilized for both "texture" (preferred grain orientation)
and residual stress measurements in the bulk of samples of interest.

     A typical instrument for wide-angle diffraction studies is
shown schematically in Figure 7. The utilization of the penetrating
power of the neutron with this type of instrument is illustrated in
Figure 8. The heavy lines parallel to "source" and "detector"
directions, respectively, represent collimators which define in-
going and scattered neutron beams. To first order, neutrons
arriving at the detector can only come from the sample volume
defined by the intersection of the projections of the collimators.
The depth of the examined volume is varied by changing the position
of the collimators with respect to the surface, but maintaining the
same scattering angle (dashed schematic). The depth to which the
neutrons might probe is dependent on the material. At approximately
1.2 Å wavelength, 50% transmission occurs for copper at ~1.1 cm
thickness (for x-rays, at ~0.001 cm thickness).

     The utility of the above-described approach has been demon-
strated for texture determination as a function depth for a more
difficult application than that of a large sample. Specifically,
the aim has been to establish the correlation between liner grain
orientation and performance in metallurgically compensated shaped-
charge munitions.

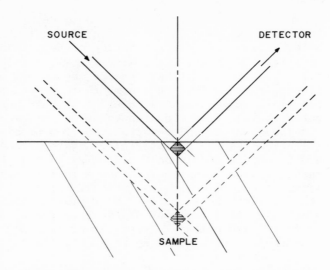

Fig. 8. Idealized geometry for texture or residual stress vs. depth
determination by means of neutron diffraction.

It has been known for many years that the optimum functioning
of certain shaped-charge ammunition requires that the projectile
have essentially zero spin (i.e., axial rotation) at impact with
the target. In contrast, spin is necessary for in-flight stability
and accuracy for many types of rounds. In spite of moderate to low
spin, it has been shown that degradation in performance can be
lessened if the shaped-charge liner is manufactured in a special
way. That is, the crystallites of the liner are aligned (by, e.g.,
shear spinning) such that as the liner collapses along preferred
slippage directions a counter-rotation equal in magnitude to that
of the projectile is thus imparted to the explosive blast jet.

Although x-ray diffraction has been used for "sampling"
inspection of metallurgically compensated liners for many years,
this technique requires (destructive) machining and etching so that
no liner for which grain orientation is known has ever been test
fired. Neutron diffraction offered the possibility of determining
grain orientation gradients throughout the bulk of a liner nondestructively.

Because the liner is only 0.28 cm thick, a "difference" technique has been developed which utilizes the fact that for a given
crystallographic plane, and appropriate masking, different volumes
(depths) are sampled when different neutron wavelengths are employed.
Alternatively, the same effect is achieved at a fixed wavelength by
examining different order reflections for a given plane.[13] This is
illustrated in Figure 9.

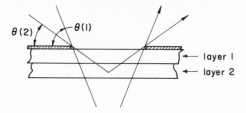

Fig. 9.   Schematic of exposure volume vs. Bragg angle for thin
          sample texture study.

The method has been used to measure texture in a test specimen
consisting of two copper plates with identical texture but mis-
aligned by 45°. In this case (200) and (400) reflections for fixed
wavelength (1.23 Å) were used. In Figure 10 are shown pole-density
distributions* for the two plates together, one plate alone, and
the individual plate pole-density distributions extracted[13] from
the two-plate data. Results are sufficiently encouraging to suggest
that, with care, texture gradients in the shaped-charge liner case
could be obtained nondestructively.

The different scattering selectivity of neutrons with respect
to x-rays is also illustrated by an example with military importance;
specifically, the determination of texture in cold-worked, tungsten-
alloy samples employed in certain antitank munitions. The samples
are composed of tungsten, nickel, and iron in a wt % ratio of
97/1.8/1.2. Sample preparation - which involves isostatic pressing,
sintering, heat treatment, and cold working - leads to a system
consisting of a "tungsten" phase (96.5 wt %) and a "matrix" phase
(3.5 wt %). The approximate compositions by weight of the two
phases are 99.7W - 0.1N - 0.2Fe and 55Ni - 23Fe -22W, respectively.
The potential of neutron diffraction for studies of this type of
sample can be inferred from Figure 11 in which partial x-ray and
neutron diffraction patterns which cover the same range in d-spacing
are shown. The full diffraction patterns show that the tungsten
phase can be indexed according to the bcc tungsten structure with

_____

*A pole figure is a map of the statistical distribution of the
 normals to specific (h k 1) planes in a sample. The counter is
 fixed at 2θ(h k 1) and the sample rotated about two perpendicular
 axes over as large an angular range as possible. Measured
 intensities reflect the distribution of normals.

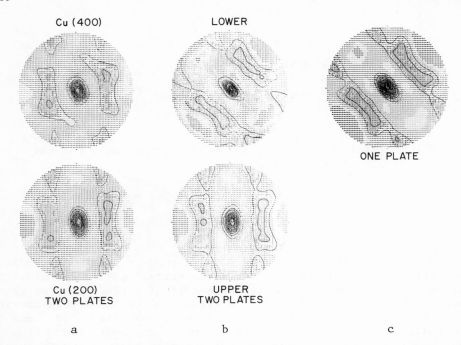

Fig. 10.   Pole-density distribution obtained from the "two-layer"
            copper sample, for α ranging from 90 to 30°.   Solid
            contour lines are drawn in multiples of random:
            (a) represents two-plate combined textures as observed by
            200 and 400 reflections; (b) shows the texture of each
            plate obtained from the bulk textures; (c) shows the
            texture of the bottom plate measured separately.

a = 3.160(1) A, and the matrix phase can be indexed according to a
fcc nickel structure with a = 3.586(1) A.   For alloys of these
compositions and relative concentrations, Bragg peaks corresponding
to the matrix phase are virtually invisible to x-rays.   In contrast,
two matrix-phase peaks are clearly visible in the neutron pattern
which makes possible texture studies and, in principal, residual
stress studies of the matrix phase.

     Texture resulting from various types of cold-working have been
measured for these types of alloys.   In Figure 12 are shown pole-
figure patterns for one matrix-phase and two tungsten-phase reflections.   The three sets of pole figures correspond to cylindrical
samples which have been cold-worked by "upsetting" with an 8%
increase in cross-section (5904), and swaged with an 8% decrease in
cross-section (5903 and 5905).   Cold-working of the latter pair was
performed at two different fabrication facilities.   As one would
expect, the fiber textures of the "upset" and swaged samples are

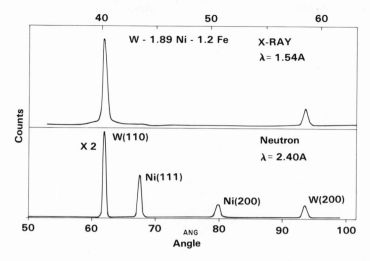

Fig. 11.   Partial x-ray and neutron diffraction patterns for a two-
          phase tungsten-alloy system.

different. On the other hand, the pattern of sample 5903 is
essentially untextured whereas that of 5905 shows definite texture
in the tungsten phase for nominally, the same cold-working.   The
matrix phase (111) pattern shows little if any clear texture.

     In contrast to the above results, samples which have been
swaged to a 16% reduction in area show very pronounced tungsten and
matrix phase textures. As shown in Figure 13, the matrix-phase
(111) pattern exhibits a more pronounced fiber texture than the
tungsten-phase (110) pattern.

     In summary, the neutron diffraction study of texture in these
samples reveals some differences resulting from cold-working that
would not have been observable with x-ray diffraction alone.
Although the significance of the measurements as they relate to
ductility, hardness, etc. is not determined, it is clear that the
nondestructive nature of the neutron probe allows empirical correlations to be made between measured properties and performance.

     Finally, a few words can be said about the possibility of
determining residual stress gradients in the bulk by means of
neutron diffraction.   Work to date in this area has focused on
demonstrating that precision and reproducibility of the measurements
are sufficient for residual stress characterization.   In Figure 14
are shown strain vs. stress results for a 1.59 cm steel rod (AISI
1017) stressed in tension.[14]   The Poisson's ratio/elastic modulus

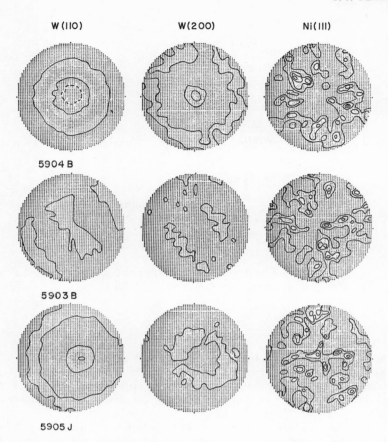

Fig. 12. Neutron diffraction pole figures for three .97W – .018Ni –
.012Fe penetrators. The samples were 0.7 cm in diameter,
10 cm in length. The center of each figure corresponds to
$\vec{Q}$ at 70° to the axis.

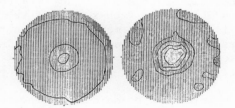

Fig. 13. Textures for a 16% R.A. swaged sample. The center point
corresponds to $\vec{Q}$ parallel to cylinder axis, the periphery
to $\vec{Q}$ at 70° to cylinder axis. W(110) is on the left,
Ni(111) on the right.

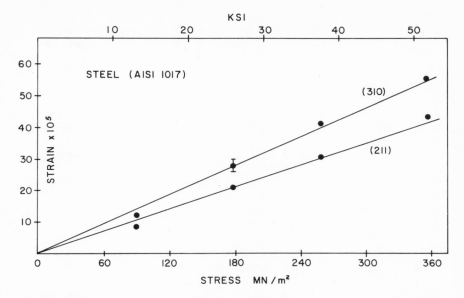

Fig. 14.   Strain vs. stress for 1.6 cm diameter steel cylinder.
Change in d-spacing measured radially for sample in
tension along cylinder axis.

(ν/E) values obtained from the slopes of the data are in excellent
agreement with theoretical predictions[15] for these planes for a
"plain carbon steel".

CONCLUDING REMARKS

The examples discussed here represent only a small sampling of
how the inherently nondestructive techniques of thermal neutron
scattering are being adapted to practical problems in the understanding of materials properties. The high penetration and selective
scattering properties of neutrons provide a powerful, and often
unique, capability to study, for example, changes in the microstructure of bulk specimens maintained in extreme environments.
The principal drawback of the neutron methods is, of course, that
the required fluxes are at present available only from research
reactors. Thus the methods are limited to the study of prototypes
rather than production-line or in-field examination.

With regard to SANS, the first state-of-the art facilities in
the United States are just now coming into use. The new instruments
at the National Bureau of Standards, Oak Ridge National Laboratory,
and the University of Missouri, for example, are being planned to
provide wide access to the scientific community. The advanced

instrumentation at these centers, in particular the use of area detectors, will significantly shorten the time required to make measurements. Hence a broader utilization and a widening range of applications for SANS can be forseen in the years ahead.

REFERENCES

1. G. E. Bacon, "Neutron Diffraction," Oxford Press, London (1977).
2. A. Guinier and G. Fournet, "Small-Angle Scattering of X-rays," Wiley, New York (1955).
3. W. Schmatz, T. Springer, J. Schelten, and K. Ibel, J. Appl. Cryst. 7:96 (1974).
4. J. Schelten and R. W. Hendricks, J. Appl. Cryst. 11:297 (1978).
5. C. J. Borkowski and M. K. Kopp, Rev. Sci. Instrum. 46:951 (1975).
6. P. Cortese, P. Pizzi, H. Walther, G. Bernardini, and A. Olivi, Mats. Sci. and Eng. 36:81 (1978).
7. F. Frisius and M. Naraghi, Atomkernenergie 29:139 (1977).
8. R. W. Hendricks, J. Schelten, and W. Schmatz, Philos. Mag. 30:819 (1974).
9. V. Gerold and G. Kostorz, J. Appl. Cryst. 11:376 (1978).
10. G. Kostorz, Chapter V, in:"A Treatise on Materials Science and Technology," H. Herman, ed., Academic Press, New York (1978).
11. M. Fatemi (private communication).
12. H. Alperin (private communication).
13. C. S. Choi, H. J. Prask, and S. F. Trevino, J. Appl. Cryst. 12:327 (1979).
14. C. S. Choi, H. J. Prask, S. F. Trevino, H. A. Alperin, and C. Bechtold, NBS Tech. Note 995:34 (1979).
15. E. Kroner, Z. Physik 151:504 (1958); F. Bollenrath, V. Hauk, and E. H. Müller, Z. Mettallkd. 58:76 (1967).

DETERMINATION OF FUNDAMENTAL ACOUSTIC

EMISSION SIGNAL CHARACTERISTICS

Richard Weisinger

Materials Science Department
The Johns Hopkins University
Baltimore, Maryland 21218

ABSTRACT

Techniques for solving wave propagation problems
with potential uses in the area of material characterization, transducer calibration, and acoustic emission
analysis are presented. The methods discussed include
analytic integral transforms, normal modes, and finite
elements. Synthetic waveform time histories are presented which illustrate each of the methods.

INTRODUCTION

The sudden cracking of a material causes bursts of
acoustic energy to be emitted. This energy, propagating
through the material, causes the particles of the medium
to move in a path which seldom resembles the original
time history at the source. Mathematical descriptions
have been very successful in predicting the displacement of small amplitude waves, showing that the motion
is dependent on the geometry, density and elastic parameters of the material; the spatial and time characteristics of the source; and the location of the receiver
with respect to the source. Analysis of physical measurements must further consider receiver resonances and
the fact that the motion at a single point can not be
known exactly since the receiver's finite size produces
an integrated response.

This paper summarizes some of the theoretical
methods of determining the particle motion caused by a

165

known force in a well characterized medium. By varying
the source and material parameters of the model, one can
identify those parameters which cause significant variation in wave propagation and also those parameters, if
any, to which the wave is relatively insensitive. Such
theoretical results could be useful in characterizing
materials from experimentally obtained waveforms if the
applied force is known. It will be shown in the development of the theory that if the material parameters of a
half-space are known and one can assume that the source
is located at a specified point in it, then the inverse
problem of finding the time history from a given waveform can also be performed.

At the present time, it is not possible to obtain
closed form solutions for the particle displacement in
an arbitrary geometry. (Results from exact solutions
for a half-space and normal mode solutions for a sphere
will be given here.) For solids, in general, one is
limited to geometries which have no abrupt edges because
the additional restrictions imposed by the boundary conditions lead to mathematical equations which are intractable. However, exact or near-exact solutions for
many types of loading in half-space and plate problems
are known. Exact solutions[1,2] in the form of normal
mode analysis for spherical and near-spherical layered
and homogeneous geometries are also known, but this type
of solution is in the form of an infinite summation which
must be truncated in any numerical evaluation. Recently,
Chapman[3] has developed a technique which uses the ray
approximation with an asymptotic evaluation of integrals
that can economically compute accurate displacement
records in spheres. The normal mode method is not as
sensitive to high frequency body waves as techniques
such as Chapman's which, in analogy to optics, reduce a
wave propagation problem to a geometrical problem of rays
generated from a source which subsequently are reflected
or refracted upon their incidence on a boundary. However,
as presented below, satisfactory results were obtained
with the normal mode method, and body waves were easily
observed in synthetic records at times which correspond
well with arrival times computed by ray geometry. One
particular advantage of the normal mode approach is that
it is no more difficult to generate long-term records
than it is for short-term records. This is not the case
for methods which use the ray approximation.

One other technique which is gaining increasing
popularity in solving wave propagation problems is the
finite element method. This method considers the medium

as a finite number of discrete elements, interconnected
by springs with spring constants determined from material
properties. Solutions are obtained by solving large
systems of coupled equations which describe the interaction between the elements via the springs. This method
requires the use of a large computer facility to solve
the equations, but as the capabilities of computers grow,
so does the attractiveness of the method.

EXACT METHODS

The exact evaluation of synthetic time histories
is accomplished by writing down the partial differential
equations for the problem with boundary and source conditions, properly performing a sequence of integral
transforms to isolate the displacement terms on one side
of the equations, and then evaluating the inverse of the
resulting equations. The method sounds quite simple,
however, finding the appropriate inverse transforms of
the resulting equations has posed serious mathematical
difficulties. To demonstrate the use of these techniques,
we consider a simple three-dimensional axisymmetric problem with a surface loading (Fig. 1). In this case, the
displacement field U which consists of the components
(U_r, U_θ, U_z) reduces to a problem independent of U_θ. The
two remaining displacement components can be written as

$$U_r = \frac{\partial \phi}{\partial r} + \frac{\partial \chi}{\partial r \partial z} \tag{1}$$

$$U_z = \frac{\partial \phi}{\partial z} - \frac{\partial^2 \chi}{\partial r^2} - \frac{1}{r} \frac{\partial \chi}{\partial r} \tag{2}$$

where $\phi(r,z,t)$ and $\chi(r,z,t)$ satisfy the wave equations

$$c_1^2 \nabla^2 \phi = \ddot{\phi} \tag{3}$$

$$c_2^2 \nabla^2 \chi = \ddot{\chi} \tag{4}$$

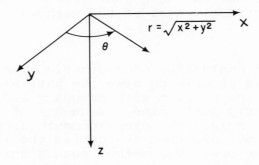

Fig. 1.   Coordinates used for a vertical force on the
          surface of a half-space.

and $\nabla^2 \equiv \dfrac{\partial^2}{\partial r^2} + \dfrac{1}{r}\dfrac{\partial}{\partial r} + \dfrac{\partial^2}{\partial z^2}$ (5)

where c_1 and c_2 are the compressional and shear wave
velocities, and the dots in Eqns. (4) and (5) refer to
partial time ($\frac{\partial}{\partial t}$) differentiation. With Eqn. (4),
Eqn. (2) can be rewritten in terms of partial derivatives
with respect to z only:

$$U_z = \dfrac{\partial \phi}{\partial z} + \dfrac{\partial^2 \chi}{\partial z^2} - \dfrac{\ddot{\chi}}{c_2}$$ (6)

We now begin the application of integral transforms.
The one-sided Laplace transforms of ϕ and χ are given as

$$\mathcal{L}(\phi) = \bar{\phi}(r,z,p) = p\int_0^\infty \phi(r,z,t)\, e^{-pt}\, dt \tag{7}$$

$$\mathcal{L}(\chi) = \bar{\chi}(r,z,p) = p\int_0^\infty \chi(r,z,t)\, e^{-pt}\, dt \tag{8}$$

where we follow Cagniard[4] in defining the Laplace transform with a p outside of the integrand. Eqns. (3) and (4) now become

$$\nabla^2 \bar{\phi} - K_1^2\, \bar{\phi} = 0 \tag{9}$$

$$K_i = \frac{p}{c_i} \qquad i = 1,2$$

$$\nabla^2 \bar{\chi} - K_2^2\, \bar{\chi} = 0 \tag{10}$$

which can be solved by the separation of variables. We define

$$\bar{\phi} = P_1(r)\, F_1(z) \tag{11}$$

$$\bar{\chi} = P_2(r)\, F_2(z) \tag{12}$$

which when substituted into Eqns. (9) and (10) yields

$$\bar{\phi} = \left[A_1 J_0(\xi r) + A_2 Y_0(\xi r)\right]\left[C_1 e^{-K_2\alpha_1 z} + C_2 e^{K_2\alpha_1 z}\right] \tag{13}$$

$$\bar{\chi} = \left[B_1 J_0(\xi r) + B_2 Y_0(\xi r)\right]\left[D_1 e^{-K_2\alpha_2 z} + D_2 e^{K_2\alpha_2 z}\right] \tag{14}$$

where $\qquad K_2\alpha_i = \left(\xi^2 + K_i^2\right)^{1/2}$ \hfill (15)

and J_0 and Y_0 are Bessel functions of the first and second kind of zero order. To obtain a physically realistic solution, we observe that $Y_0(0) = -\infty$, so that $A_2 = B_2 = 0$, and to keep the exponentials in Eqn. (15) finite, $C_2 = D_2 = 0$, provided that $\mathrm{Re}\,(K_2\alpha_i) > 0$.

A general solution to the wave equations can be found by summing over all solutions $\bar{\phi}$ and $\bar{\chi}$ given by Eqns. (13) and (14) with different values of the parameter ξ. Thus, the general solution is of the form

$$\overline{\Phi} = \int_0^\infty A(\xi, p) \, e^{-K_2 \alpha_1 z} \, J_0(\xi r) \, d\xi \tag{16}$$

$$\overline{\chi} = \int_0^\infty B(\xi, p) \, e^{-K_2 \alpha_2 z} \, J_0(\xi r) \, d\xi \tag{17}$$

When these solutions are substituted into the Laplace transforms of the displacement components, we obtain

$$u_r = \frac{\partial \overline{\Phi}}{\partial r} + \frac{\partial \overline{\chi}}{\partial r \partial z}$$

$$= \int_0^\infty [-\xi A e^{-K_2 \alpha_1 z} + \xi K_2 \alpha_2 B e^{-K_2 \alpha_2 z}] J_1(\xi r) d\xi \tag{18}$$

and

$$u_z = \frac{\partial \overline{\Phi}}{\partial z} + \frac{\partial^2 \overline{\chi}}{\partial z^2} - \overline{\chi}$$

$$= \int_0^\infty [-K_2 \alpha_1 A e^{-K_2 \alpha_1 z} + \xi^2 B e^{-K_2 \alpha_2 z}] J_0(\xi r) d\xi \tag{19}$$

where the relationship

$$\frac{\partial}{\partial r} J_0(\xi r) = -\xi J_1(\xi r) \tag{20}$$

has been used.

The Laplace transform of the relevant components of the stress tensor needed to describe a surface load are

$$\overline{\sigma}_{rz}(r, z, p) = \mu \left(\frac{\partial \overline{u}_z}{\partial r} + \frac{\partial \overline{u}_r}{\partial z} \right) \tag{21}$$

$$\overline{\sigma}_{zz}(r, z, p) = \frac{\lambda}{r} \frac{\partial(r \overline{u}_r)}{\partial r} + (\lambda + 2\mu) \frac{\partial \overline{u}_z}{\partial z} \tag{22}$$

At the surface, $z = 0$, and we define

$$\overline{\sigma}_1(r, p) = -\overline{\sigma}_{rz}(r, 0, p) \tag{23}$$

$$\bar{\sigma}_2(r,p) = -\bar{\sigma}_{zz}(r,0,p) \tag{23}$$

Substituting Eqns. (23) and (24) into the Laplace transform of the wave equations, we obtain

$$\mu \int_0^\infty [2K_2 \alpha_1 A - (2\xi^2 + K_2^2)B] \xi J_1(\xi r) d\xi = -\bar{\sigma}_1(r,p) \tag{25}$$

$$\mu \int_0^\infty [(2\xi^2 + K_2^2)A - 2\xi^2 K_2 \alpha_2 B] J_0(\xi r) d\xi = -\bar{\sigma}_2(r,p) \tag{26}$$

The form of these equations suggests that we write $\bar{\sigma}_1$ and $\bar{\sigma}_2$ as Fourier-Bessel integrals

$$\bar{\sigma}_1(r,p) = \int_0^\infty \hat{\sigma}_1(\xi,p) J_1(\xi r) \xi d\xi \tag{27}$$

$$\bar{\sigma}_2(r,p) = \int_0^\infty \hat{\sigma}_2(\xi,p) J_0(\xi r) \xi d\xi \tag{28}$$

The inverse expressions for $\bar{\sigma}_1$ and $\bar{\sigma}_2$ are

$$\hat{\sigma}_1(\xi,p) = \int_0^\infty \bar{\sigma}_1(r,p) J_1(\xi r) r \, dr \tag{29}$$

$$\hat{\sigma}_2(\xi,p) = \int_0^\infty \bar{\sigma}_2(r,p) J_0(\xi r) r \, dr \tag{30}$$

When Eqns. (27) and (28) are substituted into Eqns. (21) and (22), the unknowns A and B can be solved for as

$$A = \frac{-\xi}{\mu R(\xi)} \left[(2\xi^2 + K_2^2)\hat{\sigma}_2 - 2K_2 \alpha_2 \xi \hat{\sigma}_1 \right] \tag{31}$$

$$B = \frac{-1}{\mu R(\xi)} \left[2\xi \alpha_1 K_2 \hat{\sigma}_2 - (2\xi^2 + K_2^2)\hat{\sigma}_1 \right] \tag{32}$$

where $R(\xi)$ is the Rayleigh function

$$R(\xi) = (2\xi^2 + K_2^2)^2 - 4K_2^2 \alpha_1 \alpha_2 \xi^2 \tag{33}$$

The problem now reduces to finding the inverse

Laplace transform of the integrals in Eqns. (18) and (20) with the constants A and B as given in Eqns. (31) and (32). The effect of the Laplace transform variable p, which is contained in K_2, can be more clearly expressed by making the substitutions

$$K_1 = \gamma K_2 , \quad \xi = K_2 x , \quad d\xi = K_2 dx \tag{34}$$

where
$$\gamma = \frac{c_2}{c_1} \tag{35}$$

Equations (18), (20), (31), (32), and (33) now become

$$\bar{u}_r = K_2^2 \int_0^\infty [-Ae^{-K_2\alpha_1 z} + K_2\alpha_2 Be^{-K_2\alpha_2 z}] \times J_1(K_2 rx) dx \tag{36}$$

$$\bar{u}_z = K_2^2 \int_0^\infty [-\alpha_1 Ae^{-K_2\alpha_1 z} + K_2 x^2 Be^{-K_2\alpha_2 z}] J_0(K_2 rx) dx \tag{37}$$

$$A = \frac{-x}{K_2 \mu R(x)} [(2x^2+1)\hat{\sigma}_2 - 2\alpha_2 x \hat{\sigma}_1] \tag{38}$$

$$B = \frac{-1}{K_2^2 \mu R(x)} [2\alpha_1 x \hat{\sigma}_2 - (2x^2+1)\hat{\sigma}_1] \tag{39}$$

$$R(x) = (2x^2+1)^2 - 4x^2 \alpha_1 \alpha_2 \tag{40}$$

where now
$$\alpha_1 = (x^2 + \gamma^2)^{1/2} \tag{41}$$

and
$$\alpha_2 = (x^2 + 1)^{1/2}$$

Under the integrals, the Laplace transform variable appears as parameters of the Bessel Functions, exponentials, and of the stress components, and outside the integral, it appears raised to the first power. We note that the source is a casual impulse and thus, the response must also be casual. This means that the Laplace transform, as applied above, is equivalent to the Fourier transform with the change of variable p = iw.

It is often also possible to separate the spatial from the time parts of the source.

A very simple example is that of a verticle point source with step function time dependence. In this case

$$\sigma_2(r,t) = -F\,H(t)\,\frac{\delta(r)}{\pi r} \tag{42}$$

$$\sigma_1(r,t) = 0 \tag{43}$$

and the transformed counterpart of σ_2 is

$$\hat{\sigma}_2(r,p) = \frac{-F}{\pi}\,\mathcal{H}\left(\frac{\delta(r)}{r}\right)\mathcal{L}\left(H(t)\right) = \frac{-F}{2\pi} \tag{44}$$

where we use the $\mathcal{H}$ to refer to the Hankel transform, $H(t)$ is a step function, and $\delta(r)$ is the spike delta function. In this special case of a step function, the Laplace frequency variable p does not appear in the integral. For more complicated time variations, an expression involving the p's will appear, but if it is not coupled to the spatial components, it can be simply taken out from under the integral sign.

It is now observed that if one wishes to experimentally determine time characteristics of arbitrary point sources one first obtains a frequency spectrum of a signal resulting from a step function point source, and equates this experimentally measured spectrum with the integrals of Eqns. (18) and (20) ranging over all frequencies. Then the frequency spectrum of any arbitrary signal generated from a known point can then be "divided" by the response for a step function source. This can then be inverted into the time domain to give a representation of the time history of the source. Hsu and Hardy[5,6] have previously described a deconvolution technique which uses a Green's function for a plate based on a similar idea.

For a finite-size source, the Hankel and Laplace transforms can not usually be separated as was done in Eqn. (44), and in these cases the frequency term can not be taken outside the integral.

One of the earliest and still one of the best methods for finding the inverse Laplace transform of integrals of the form of Eqns. (36) and (37) is that of Cagniard. This method avoids the entire problem by manipulating the integral into the form of a Laplace transform, and then simply removing the integrand to find the inverse Laplace transform. No actual integration needs to be performed. However, the process of changing the variables of integration often leads to lengthy algebraic manipulations and tracing integration paths in the complex plane. Dix[7] has described a simple application of the method Cagniard originally solved only a two-dimensional problem, but the method was extended by deHoop[8] to three dimensions. The Cagniard method is often referred to as a general semi-inverse method.

Abramovici[9] using the Cagniard method, has given a general solution for finding the inverse Laplace transform for equations of the form

$$K^n \int_0^\infty f(x) \, J_m^{(s)}(Krx) \, e^{-Khg(x)} \, dx \qquad (45)$$

where h and r are constants, $J_m^{(s)}$ is the s-order derivative of the Bessel function of the first kind, K is the Laplace transform variable divided by the wave velocity, f(x) is a rational function of x^2 square roots of the form $\sqrt{x^2 + \lambda^2_j}$ with λ_j being constants.

The Bateman-Pekeris Theorem[10] was developed to invert the integrals by finding the inverse Laplace transform of the integrands, and then numerically evaluating the resulting integral. This theorem overcomes the problem that an analytic expression for the inverse Laplace transform of the Bessel function of the first kind is not known. By proper manipulation, the integrand can be converted into an expression involving a modified Bessel function whose inverse transform is known. This method was later extended by Aggarwal and Ablow[11]

Longman[12,13,14] in a series of papers showed how integrals similar to those of Equations (36) and (37) could be inverted by a technique called the best rational function approximation. A series expansion to the solution is made which Longman has found to converge rapidly to the exact solution.

The first exact displacement-time history records were given by Lamb[15]. His results were later rederived with integral transform techniques similar to those outlined above by Pekeris[16]. Figure 2 shows the theoretical record of Lamb's problem for the vertical dis-

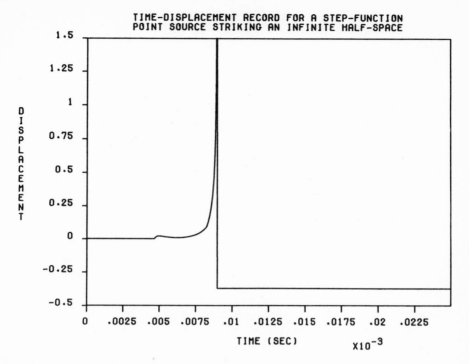

Fig. 2.   Vertical particle displacement for Lamb's
          problem at the surface.

placement received on the surface of a homogeneous
isotropic elastic half-space caused by a point load with
step function time dependence. Figure 3 shows the vertical displacement slightly below the surface of the half-
space. A clear arrival of the shear wave is now visible.
Figure 4, the antipodal solution or often misnamed epicenter solution, is the vertical displacement of a point
immediately below the surface. Chao[17] has considered
a tangential point surface force. Gakenheimer[18]
further extended the Cagniard method to include moving
surface loads, and Roy[19] extended the work of Gakenheimer to derive surface displacements in an elastic
half-space due to a buried moving point source. There
is a large literature with solutions to similar problems.

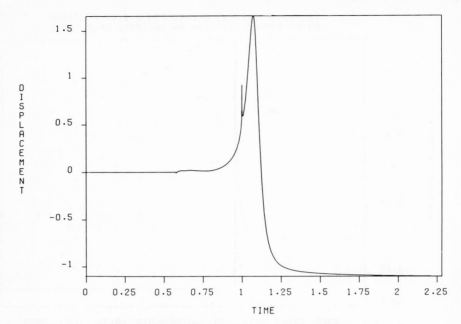

Fig. 3.   Vertical particle displacement for Lamb's
          problem just below the surface.

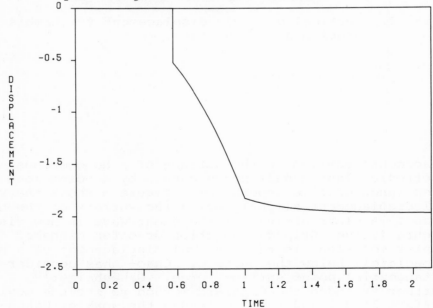

Fig. 4.   Vertical particle displacement for Lamb's
          problem immediately below the source.   (The
          antipodal solution)

NORMAL MODE SOLUTION

The normal mode solution is severely limited in its applicability. Few exact analytic solutions for the resonant modes of vibration of finite bodies exist. Solutions for spherical, near spherical and torroidal bodies do exist. However, the need to satisfy the boundary conditions for bodies of other geometries makes a mathematical solution intractable. Perhaps in the future, with the further development of numerical Rayleigh-Ritz procedures, solutions for geometries of arbitrary shape will be obtained.

The method consists of finding the resonant frequencies for the body by solving the equations of motion with the appropriate boundary conditions. Once these are known, one can determine the shape of the motion which occurs from the resonant vibration at that frequency. Motion of the body at the lowest frequencies of resonance are the simplest, and become progressively complex at higher frequencies. Figures 5 and 6 show some of the mode shapes of a sphere. In Fig. 4, the $_1T_2$ torroidal mode is an overtone of the $_0T_2$ mode, the former performing the same motion as the latter, but on two different levels. The shapes of Fig. 6 have been termed the "breathing mode" ($_0P_0$), the "football mode" ($_0P_2$), and the "cloverleaf mode" ($_0P_3$).

A source acting on the sphere will cause each of the different normal modes to become active, each contributing to the total particle displacement in the body. The approach can be thought of as a means of determining the time domain from the frequency domain.

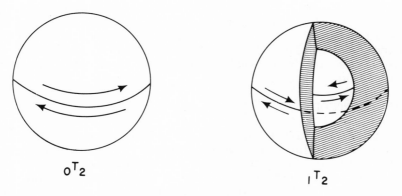

$_0T_2$ $_1T_2$

Fig. 5.   Torroidal mode shapes for a sphere.

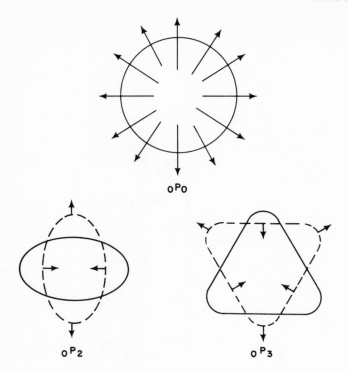

Fig. 6.   Poloidal mode shapes for a sphere.

For a sphere with an axisymmetric source, only
one parameter θ is needed to specify a position on
the sphere's surface. Figure 7 shows how θ is defined.
Figures 8 and 9 show the particle displacement at $\theta = 20°$
and 80° caused by a point source located at $\theta = 0°$ with
a step function time dependence. Note the similarity
to Lamb's problem of Fig. 2 in the initial arrivals on
the $\theta = 20°$ plot. The letters P, S and R refer to the
phase arrivals of the compressional, shear, and Rayleigh
waves, respectively. The lower frequency Rayleigh
surface waves are much more dominant than the higher
frequency body waves.

NUMERICAL METHODS

Of the numerical methods, finite element and finite
difference methods[20] are most common. With the increasing availability of large computers, these methods
will undoubtedly become much more widespread, and their
increased use will probably also lead to improvements
in their efficiency. Many of the codes now available

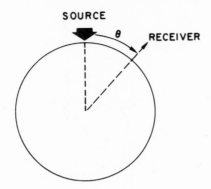

Fig. 7.   θ, angle of colatitude for a spherical
           geometry.

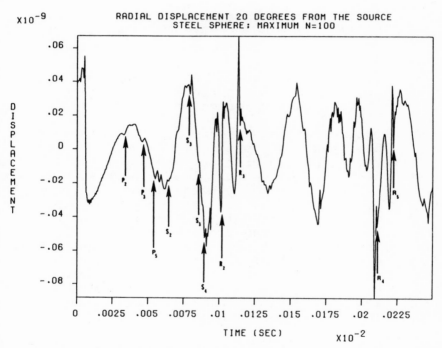

Fig. 8.   Particle displacement at θ = 20° on a sphere
           caused by a point surface force acting at
           θ = 0° with step-function time dependence.

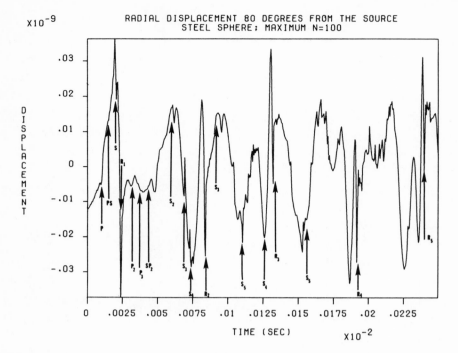

Fig. 9.   Particle displacement at θ = 80° on a sphere
           caused by a point surface force acting at
           θ = 0° with step-function time dependence.

for dynamic time analysis treat only the axisymmetric
case, as this involves essentially only a two-dimensional statement of the problem. Any symmetry leads to
increased savings in computer time (= money). There are
also codes which treat the general dynamic three-dimensional case, however, most of these codes were not
developed explicitly for solving wave propagation
problems, because they require a very fine mesh, and the
introduction of an additional dimension leads to a large
increase in the number of elements needed for the simu-
lation of the problem. The vertical displacement for
Lamb's problem solved for an 81 x 81 mesh with an
axisymmetric dynamic finite element code is shown in
Fig. 10. This should be compared with Fig. 2.

MATERIAL CHARACTERIZATION

     Variations of the received signal with direction
can give a representation of the surface texture of a
material. Isotropic materials with different material
properties will transmit signals from identical sources

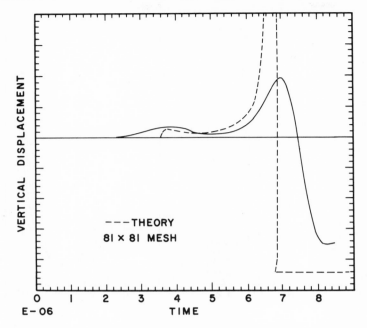

Fig. 10.   Finite element computation for vertical dis-
           placement of Lamb's problem on a 81 x 81 mesh
           compared with integral transform theory.

with different results.   Perhaps in the future a characterization of material surface anisotropy will be
possible by observing the change in received signals
from a given source at different locations.

     The variation of material parameters has a notice-
able effect on the structure of the received signal.
For example, Mooney[21] considered the variation of the
signal due to the variation of the Poisson ratio σ.
Those results are shown in Fig. 11.   Noticeable differences in the shape arrival times and final offset of the
displacement are evident.   This difference in final offset of the displacement are evident.   This difference
in final offsets suggests an application.   From the clean
experimental results presented by Breckenridge[22],
shown in Fig. 12, it would seem that it may be possible
to evaluate the Poisson ratio of a material from a
measurement of the difference between the final offset
of the signal with the original zero signal baseline; or
also equally possible, from the ratio of the shear and
longitudinal velocities, given in terms of the Poisson
ratio as

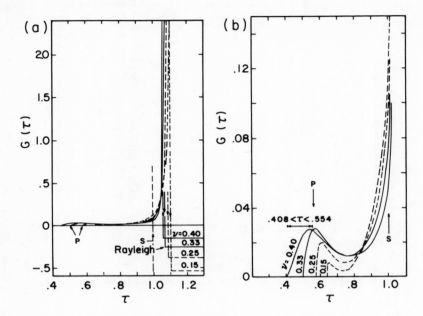

Fig. 11.   (a) Effect of Poisson's ratio on the observed
           wave forms for a step-function source;   (b)
           expanded 20 times.   (After Mooney[21])

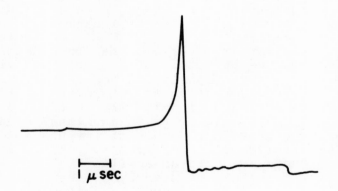

Fig. 12.   The experimental surface vertical displacement
           caused by a vertical step-function load on the
           surface of a large flat block before any
           reflections occur.   (After Breckenridge[22])

$$\left(\frac{c_1}{c_2}\right)^2 = \frac{2(1-\sigma)}{1-2\sigma} \tag{46}$$

where the shear velocity can be determined by measuring the Rayleigh wave velocity and noting that

$$c_2 = \frac{c_R}{\sqrt{\gamma_1}} \tag{47}$$

where c_R is the Rayleigh velocity and γ_1 is the root of the Rayleight function of Eqn. (33). Bell[23,24] has pointed out that easy and reliable measurements of the Poisson ratio have always been difficult to make. Once the density and wave velocities are known, the Lamé constants and Young's modulus can easily be calculated.

SUMMARY

In this paper, methods of theoretically computing time-displacement histories of particle motion are given. The integral transform methods give the most desirable solutions because they are exact or near exact. There are often, however, mathematical complexities which result in the final integral inversions needed to determine displacement. A variety of methods have been proposed for overcoming such difficulties. Those outlined here were the method of Cagniard, Bateman-Pekeris, and Longman. The applicability of normal mode solutions is much more limited than the transform techniques, but in suitable geometries, such as spherical or near-spherical, long-term records with the appearance of many wave reflections can be easily obtained. Similar long-term records obtained from transform methods would require lengthy computations in determining the contributions from the ray arrivals due to reflections. Perhaps the easiest generation of records can be achieved through the application of large-scale finite element computer programs. The greatest advantage of finite elements is that virtually any type of geometrical structure can be modeled, whereas exact solutions fail for all but the simplest geometries. For a good wave representation, rather than a crude approximation, enormous amounts of computer time and memory are required for dynamic finite element wave propagation solutions. However, as computer costs decrease, and both the computers and the finite element codes become increasingly sophisticated, the numerical approach will become more attractive.

Theoretically predicted time-displacement records can yield important information about the time and spatial history of the source as well as material characteristics. Further work in the theoretical calculation of wave forms should more clearly define acoustic emission signal characteristics and can further aid in the interpretation of experimentally observed waveforms.

REFERENCES

1. M. Saito, Excitation of Free Oscillations and Surface Waves by a Point Source in a Vertically Heterogeneous Earth, J. Geophys. Research, 72:3689-3699 (1967).

2. R. Weisinger, Acoustic Emission and Lamb's Problem on a Homogeneous Sphere, Masters' Essay, The Johns Hopkins University, Baltimore, Maryland (1979).

3. C. H. Chapman, A New Method for Computing Synthetic Seismograms, Geophys. J. R. Astron. Soc., 54:481-518 (1978).

4. L. Cagniard, "Reflection and Refraction of Progressive Seismic Waves," trans. Dix, C. H. and Flinn, E. A., McGraw-Hill Book Company, New York (1962).

5. N. N. Hsu, J. A. Simmons and S. C. Hardy, Acoustic Emission Signal Analysis--Theory and Experiment, Materials Evaluation, 35: 100-106 (1977).

6. N. N. Hsu and S. C. Hardy, Experiments in Acoustic Emission Analysis for Characterization of AE Sources, Sensors and Structures, Winter Meeting ASME, "Elastic Waves and Non-destructive Testing of Materials," AMD-Vol. 29 (1978).

7. C. H. Dix, The Method of Cagniard in Seismic Pulse Problems, Geophysics, 19:722-738 (1954)

8. A. T. deHoop, A Modification of Cagniard's Methods for Solving Seismic Pulse Problems, Appl. Sci. Res., 8B:349-356 (1961).

9. F. Abramovici, A Generalization of the Cagniard Method, Journal of Computational Physics, 29:328-343 (1978).

10. H. Bateman and C. L. Pekeris, Transmission of Light from a Point Source in a Medium Bounded by Diffusely Reflecting Parallel Plane Surfaces, J. Optical Soc., 35:645-655 (1945).

11. H. R. Aggarwal and C. M. Ablow, Solution to a Class of Three-Dimensional Pulse Propagation

Problems in an Elastic Half-Space, Int. J. Engng. Sci., 5:663-679 (1967).

12. I. M. Longman, Computation of Theoretical Seismograms, Geophys. J., 21:295-305 (1970).

13. I. M. Longman, Best Rational Function Approximation for Laplace Transform Inversion, SIAM J. Math. Anal., 5:574-580 (1974).

14. I. M. Longman and T. Beer, The Solution of Theoretical Seismic Problems by Best Rational Function Approximations for Laplace Transform Inversion, Bull. Seis. Soc. Am. 65:927-935 (1975).

15. H. Lamb, On the Propagation Tremors Over the Surface of an Elastic Solid, Phil. Trans. Royal Soc. London, A203:1-42 (1904).

16. C. L. Pekeris, The Seismic Surface Pulse, Proc. Nat. Acad. Sci., 41:469-480 (1955).

17. C. C. Chao, Dynamic Response of an Elastic Half-Space to Tangential Surface Loadings, J. Appl. Mech., 27:559-567 (1960).

18. D. C. Gakenheimer, Transient Excitation of an Elastic Half-space by a Point Load Traveling on the Surface, J. Appl. Mech., 26:505-515 (1969).

19. A. Roy, Surface Displacements in an Elastic Half-Space Due to a Buried Moving Point Source, Geophys. J. R. Astron. Soc., 40: 289-304 (1974).

20. O. C. Zienkiewicz, "The Finite Element Method," third edition, McGraw-Hill Book Company, New York (1977).

21. H. M. Mooney, Some Numerical Solutions for Lamb's Problem, Bull. Seis. Soc. Am., 64: 473-491 (1974).

22. F. R. Breckenridge, C. G. Tschiegg and M. Greenspan, Acoustic Emission: Some Applications of Lamb's Problem, J. Acous. Soc. Am., 57:626-631 (1974).

23. J. F. Bell, On the History of Some Recent Measurements Described in Experimental Mechanics, 17:359-360 (1977).

24. J. F. Bell, The Experimental Foundations of Solid Mechanics,"Handbuch der Physik," VIa/1, Springer-Verlag (1973).

A SIMPLE DETERMINATION OF VON KARMAN CRITICAL VELOCITIES

R. B. Pond, Sr.* and J. M. Winter, Jr.

Marvalaud, Inc.
Westminster, Md.

ABSTRACT

The ductility of non-brittle engineering metals and alloys has been shown by Hoppmann to be a function of deformation velocity. The ductility increases to some maximum at the Karman Critical Velocity and thereafter diminishes. This implies that the volume of the plastic zone at the tip of the crack will vary as a function of deformation velocity, and that at the Karman velocity the character of the material drastically changes toward a brittle material. An analysis of this phenomena is presented. A tensile test is generally required to determine the critical value. This technique is simplified by using the non-destructive characterization of the scleroscope hardness which is shown to rank the critical velocities in a reproducible manner. The reasons for this relationship are analyzed by using the Pond - Glass method of critical velocity determination.

INTRODUCTION

The importance and value of non-destructive testing has been firmly established by recognition of the fact that although most engineering materials may be statistically uniform in their virgin state, fabrication and/or use can induce dangerous flaws which are not uniformly distributed. The principal focus of NDE today is in the location of flaws and measurement of the severity of such flaws be they blow holes, shrinkage cavities, pores, cracks, undesired second phase, or areas of high stress concentration.

---

*R. B. Pond, Sr. is also with The Johns Hopkins University, Baltimore, Md.

Stated in a slightly different way, the principal focus of NDE
today is the determination of the existence (and degree) of discontinuities and inhomogeneities which will produce an undesired
change in mechanical behavior.

The materials scientist is not only interested in mechanical
behavior but also (and most often) how it can be altered with
possible, allowable, and dominant changes in the internal structural arrangement. He must determine the level of properties
such as strength, hardness, etc. from which the flaw produces a
deviation. In order to accomplish this on explicit items of hardware a system of NDE must be utilized. Hardness testing methods
are most frequently used to evaluate these property levels. The
question arises, "Since all hardness testing involves some degree
of surface penetration, is hardness testing truly non-destructive?".
The purpose of this paper is not to argue this question. The
hardness measuring technique which produces the least distortion
is the scleroscope technique. Inherent in this measurement is the
capability of indexing a value of the von Karman critical velocity.
This critical velocity should be considered in evaluating the
severity of a crack in a semi-brittle or ductile material subjected to dynamic loads.

The independent establishment of the theory of plastic strain
propagation in metallic materials, in 1941, by von Karman[1] and
Taylor[2] is nicely documented by D. S. Clark in his Campbell
Memorial Lecture[3]. The experimental verification of the existence
of the proposed critical velocity was presented by Duwez & Clark[4]
and by Hoppmann[5] in 1947. Hoppmann also showed, as did Clark,
(see Fig. 1), that at the critical velocity the ductility of the
material was maximized. This ductility enhancement was proposed
by Pond[6] as the explanation of the behavior of the metal in the
cold extruding of long, thin walled tubes and was used by Pond &
Glass[7] to establish the critical velocity value for single crystal
aluminum. The approach was also used by Pond & Glass[8] to understand cratering in hypervelocity impacts. There seems to be no
doubt that the critical velocity is a property of metallic materials and that the ductility of the non-brittle material at this
velocity is maximized.

A confusion which arises from reading these references may
come from the distinction between impact velocity and strain rate.
An easy way to visualize the difference is to note that in cold
impact extrusions, the ram velocity (impact velocity) may be
10"/sec but the flow velocity in the vicinity of the orifice (for
an extrusion ratio of 90) will be greater than 900 "/sec. If a
single crack with a tip radius of 5×10^{-4} inches exists in a
specimen which is being pulled apart at a rate of 1.0 "/sec the
material at the tip of the crack could be flowing at the rates
above 1000 "/sec. This implies that since, (per the foregoing

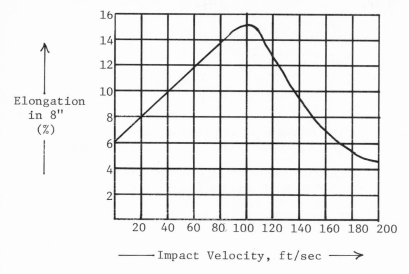

Fig. 1:   Effect of Impact Velocity on Percentage
          Elongation of SAE 1020 Steel, Cold Rolled

discussion), ductility is a function of velocity, the ductility
of the material at the crack tip will vary with 1) the deformation
rate of the part, 2) the radius of the crack tip and 3) the material involved. Since the ductility of the material decreases with
velocity above the critical velocity, it approaches the state of
a brittle material above this value.

        The critical velocity therefore indicates the velocity above
which the structure will behave as a brittle one and below which
cracks should blunt or become less effective due to larger zones
of plastic strain with corresponding higher energy absorption at
the crack tip.

        For these reasons it becomes important to know the Karman
Critical Velocity (KCV) for a material. Since the KCV can be
changed by local reactions such as work hardening, work softening,
precipitation, etc. it would be advantageous to be able to measure
the KCV on a local scale. In normal practice this is not possible
since the value is determined from a stress-strain curve of the
material which has been extended to rupture.

DEVELOPMENT OF KCV

        In the theory of plastic wave propagation developed by both
von Karman[1] and Taylor[2], the relation between the plastic strain
(ϵ) and impact velocity (V) is:

$$V(\epsilon_1) = \int_0^{\epsilon_1} c(\epsilon)d\epsilon \qquad \text{(Eqn. 1)}$$

where ϵ_1 is the maximum strain developed from the
striking velocity, V.

and $c(\epsilon)$ is the wave speed of the plastic (or elastic)
tensile wave of strain ϵ .

The plastic (or elastic) wave speed can be obtained from the
relation:

$$c(\epsilon) = \left[\frac{d\sigma}{d\epsilon} \bigg/ \rho_0 \right]^{1/2} \qquad \text{(Eqn. 2)}$$

where $\frac{d\sigma}{d\epsilon}$ is the slope of the static engineering
(nominal) stress-strain curve at strain
value   .

and ρ_0 is the mass density of the undeformed material.

The value of the slope of the static engineering stress-
strain curve goes to zero at the strain value ϵ (UTS) associated
with the ultimate stress, (corresponding to the onset of necking
in the tensile specimen). It is obvious that the plastic wave
speed then goes to zero at this strain, which means the integral
relation for impact velocity will produce a singular (critical)
maximum value, V_o, designated the "Karman Critical Velocity", KCV.

$$KCV = V_o = \int_0^{\epsilon_{UTS}} c(\epsilon)d\epsilon \qquad \text{(Eqn. 3)}$$

The sequence of relationships associated with this formulation
are shown schematically in Figures 2 a, b, and c.

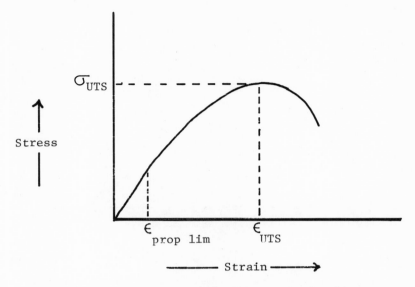

Fig. 2a: Nominal Static Stress–Strain Curve

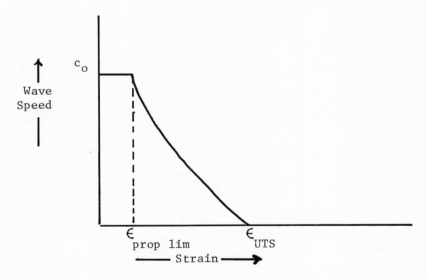

Fig. 2b: Plastic Wave Velocity as a Function of Strain,
           Calculated from Curve in Fig. 2a.

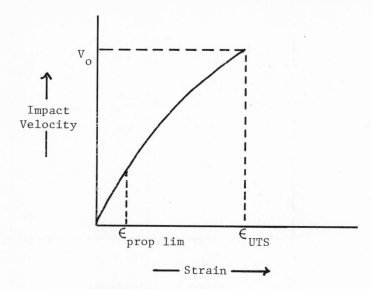

Fig. 2c: Impact Velocity as a Function of Strain,
           Calculated from Curve in Fig. 2b.

        This model, although elegant, is unsatisfactory to the
materials scientist who prefers to represent the stress-strain
relationship more descriptively as a true-stress, true-strain
curve. In the Pond-Glass model[7] the KCV is computed from the
dynamic stress-strain curve which is derived from the true-stress,
true-strain static curve. It is presumed that 1) the maximum
stress (static or dynamic) is properly displayed, 2) this stress
can be achieved elastically and 3) when this stress is elastically
achieved the material will flow as a viscous fluid, i.e., no
strain hardening will occur. This set of conditions is represented
in Fig. 3.

        For the dynamic stress-strain curve in Fig. 3, the only non-
zero wave speed of Eqn. (2) becomes the elastic wave velocity, c_o:

$$c_o = \left[E \middle/ \rho_o \right]^{1/2} \qquad\qquad \text{(Eqn. 4)}$$

        where E is Young's modulus and Eqn. (3) then can be
        taken to yield;

$$\underline{KCV} = V_o = \left[E \Big/ \rho_o \right]^{1/2} \epsilon_{elastic, \ dynamic}$$

$$= \left[\sigma_{TRS}^2 \Big/ \rho_o E \right]^{1/2} \qquad \text{(Eqn. 5)}$$

where σ_{TRS} is the true stress at rupture.

Another way to look at this result is to reason that in order to achieve the non strain hardened flow condition, it is necessary to deliver to the test specimen enough kinetic energy to just equal the dynamic elastic energy represented by the area under the elastic portion of the dynamic curve (the dynamic modulus of resilience). Thus

$$\tfrac{1}{2} \rho_o V_o^2 = \sigma_{TRS}^2 \Big/ 2E$$

or

$$\underline{KCV} = V_o = \left[\sigma_{TRS}^2 \Big/ \rho_o E \right]^{1/2} \qquad \text{(Eqn. 6)}$$

which is the same result as in Eqn. 5.

SCLEROSCOPE HARDNESS

Hardness is a most difficult word to define and penetration hardness is not easily understood. However it is obvious that the Shore scleroscope method measures something different from the numbers produced by the Rockwell & Brinell penetration methods although superficial indentations are produced by the impacting hammer of the scleroscope. The scleroscope method is a dynamic one in which a tup (or hammer) is allowed to fall a fixed distance under its own weight and to rebound (or bounce) from the test specimen surface. The height to which the tup rebounds is taken as a measure of the hardness of the material. The initial impacting area of the tup is virtually a point and since the initial stress on the material being tested is extremely high, a slight degree of plastic deformation occurs. Since the tup is spherical shape, the contact area increases rapidly with this slight deformation and the material is loaded only elastically.

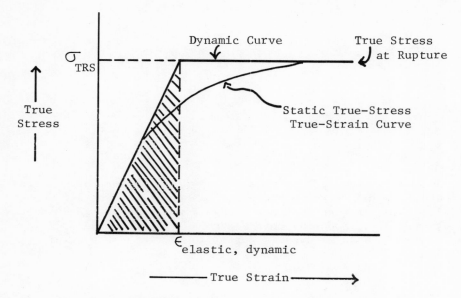

Fig. 3:   Static and Dynamic True-Stress True-Strain
          Curves

The rebounding elastic wave unloads the specimen and propels the
tup to a height dependent upon the amount of elastic energy stored
in a specimen. The amount of plastic deformation generated is not
only a function of the small contact area but is also a function
of the dynamic elastic limit.

     Since the scleroscope gives an index of the dynamic elastic
limit as well as the dynamic modulus of resilience, it seems
highly probable that its index should correlate very well with
the KCV of a material.

TEST RESULTS

     In order to test this proposition, use was made of a single
bar of OFHC copper of sufficient size to provide eight tensile
test specimens having the same dimension as one another. Such a
provision insured that the chance of chemical difference between
the test specimens was negligible and that one cross head velocity
of the testing machine would provide the same strain rate for all
the specimens.

     The original material was annealed in an inert atmosphere
@ 810°C for two hours and a tensile specimen derived therefrom.
The bar was then cold rolled and a tensile specimen taken after
each of the seven reductions; 24% (reduction of area), 44%, 75%,

86%, 93%, 96%, & 98%. The Shore scleroscope number was determined
for each of the eight specimens by averaging 16 values taken at
16 different spots on each specimen.

Tensile values of load, displacement & cross sectional area
were then obtained for each specimen using a cross head speed of
0.05 cm/min. The dimensions of each specimen were 0.33 cm. diameter x 5 cm. gauge length. A true-stress, true-strain curve was
developed for each of these specimens and (using the method previously described) the KCV was determined for each specimen.

The correlation between the scleroscope index and the KCV
for copper is illustrated in Fig. 4.

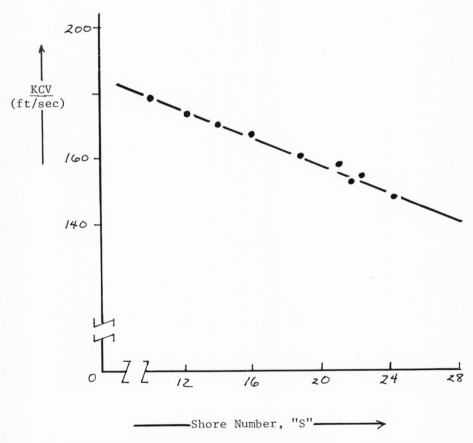

Fig. 4: Measured Shore Scleroscope Number versus Computed
KCV for Copper with Varying Degrees of Cold Work

DISCUSSION & CONCLUSIONS

It is proposed that as more is learned about the reaction of materials having internal damage it will become necessary to establish the KCV for materials being used. It is only with such knowledge, coupled with knowledge of flaw dimension and the velocity of the anticipated displacements that one will be able to classify materials as brittle or non-brittle.

Whether or not this proposition is correct, it can be concluded that KCV values in OFHC copper varied by work hardening can be determined by the Shore scleroscope test. It is interesting to speculate that perhaps KCV values for many metals can be determined by the Shore Scleroscope test. The authors intend to provide additional data for a number of other metals.

ACKNOWLEDGEMENT

The authors wish to acknowledge the original measurements made by Mr. Walter Panowicz and used in this paper.

REFERENCES

1. Th. v. Karman "On The Propagation of Plastic Strains In Solids, "Nat. Defense Research Council Rept. No. A-29 (OSRD No. 365) (1942).
2. G. I. Taylor, 6th International Congress of Applied Mechanics, Paris, (1946).
3. D. S. Clark "The Behavior of Metals under Dynamic Loading", Trans. ASM, 46:34, (1954).
4. P. E. Duwez & D. S. Clark, "An Experimental Study of the Propagation of Plastic Deformation Under Conditions of Longitudinal Impact", Proc. ASTM, 46, (1954).
5. W. H. Hoppmann, Proc. ASTM, 47:533, (1947).
6. R. B. Pond, "Cold Extrude Rapidly to Produce Long, Thin-wall Aluminum Tubes", Met. Progress 6:89, (1966).
7. R. B. Pond & C. M. Glass, "Crystallographic Aspects of High Velocity Deformation of Al-Single Crystals," Response of Metals to High Velocity Deformation, Interscience Publishers, Inc., New York, (1961).
8. R. B. Pond & C. M. Glass, "Metallurgical Observations and Energy Partitioning" High Velocity Impact Phenomena, Academic Press (1970).

FRACTURE PREDICTION BY RAYLEIGH WAVE SCATTERING MEASUREMENT

M. T. Resch, J. Tien, B. T. Khuri-Yakub,
G. S. Kino, and J. C. Shyne

Stanford University
Stanford, California 94305

ABSTRACT

An acoustic surface wave NDE technique for predicting the fracture stress of a solid containing a surface crack is described. The normalized stress intensity factor for a crack is determined by measuring the reflection coefficient of Rayleigh waves incident to the crack in the long wave-length limit. An acoustic surface wave wedge transducer was used to excite the incident wave and to measure the reflected wave intensity. The fracture stress of Pyrex glass discs containing the acoustically measured cracks was determined in biaxial flexure. Fracture toughness values obtained through post-fracture examination of the fracture surfaces were in excellent agreement with the values in the literature. The fracture stress predicted from the acoustic measurement of the stress intensity factor correlated very well with the measured value of the fracture stress. For surface cracks with crack radii varying between 50 and 500 microns, the average error between the acoustic prediction of fracture stress and the measured fracture stress is less than 20%.

INTRODUCTION

We describe here a new quantitative acoustic surface wave technique for nondestructive evaluation of the fracture stress of brittle solids containing surface flaws. The flaw type treated in this paper consists of flat, semi-elliptical shaped cracks. We shall refer to them as being roughly "half-penny shaped," since this terminology is in common usage. For the present study, Pyrex glass was chosen as the experimental material on the basis of

197

availability and its well documented fracture behavior. It has
also been shown that our measurement technique is applicable to
structural ceramic materials like silicon nitride.[1] This work
will be described in a separate paper. The procedure is to excite
a Rayleigh wave on the surface of our glass specimens and to compare the amplitude of the wave incident on the surface crack with
the amplitude of the reflected echo. These measurements are ideally
performed at a frequency such that the crack dimensions are much
smaller than the acoustic wavelength. This method of flaw evaluation is based upon long wavelength acoustic scattering principles
applicable both to interior cracks completely surrounded by bulk
solid and to surface cracks intersecting a free surface. A flat,
elliptical shaped crack will distort the dynamic stress field of
a uniform, plane longitudinal acoustic wave in exactly the same way
that it perturbs the static stress field resulting from the application of external forces to the body. It is this similarity that
allows expression of an analytical relationship between the reflection coefficient of an acoustic wave incident on the crack and the
stress intensity factor at the crack tip. Once the stress intensity factor of a crack is known, fracture stress due to the flaw
may be predicted if we know the fracture toughness of the material
in question.

When Rayleigh waves are used with surface cracks, a long
acoustic wavelength method is particularly applicable because of
the nature of the dynamic stress field associated with a Rayleigh
wave.

The amplitude of the longitudinal stress of a Rayleigh wave
diminishes exponentially with increasing depth below the surface,
scaled in proportion to the wavelength. Rayleigh waves with wavelength long relative to the crack depth can be approximated as
having uniform stress amplitude over the entire surface of the
surface crack.

THEORETICAL BACKGROUND

Budiansky and Rice [2] have demonstrated the possibility of
determining the maximum stress intensity factor of flat, elliptical
shaped cracks contained in a volume of material. Their analysis
requires determination of the reflection coefficient of bulk waves
from the crack at three different angles of incidence. We have
particularized this theory in order to evaluate the maximum stress
intensity factor for the case of surface wave scattering from
surface cracks. In this case, it is possible to determine optically the direction of the normal to the crack face, so that only
one measurement of the reflection coefficient normal to the crack
plane is needed.

Consider a flat, elliptical shaped surface crack residing in a solid with a coordinate system as shown in Fig. 1. Here a and c are the lengths of the semi-minor and semi-major axes of the half ellipse, respectively. The position s along the crack tip is defined by the angle θ. At s a unit $\vec{n}$ from the crack edge lies in the x-y plane, and infinitesimal length $\vec{dl}$ is tangent to the crack edge. Additionally, we define the distance ρ from the origin along a line perpendicular to the tangent line extending from s.

A farfield stress with component σ_{zz} normal to this crack will cause a stress σ near the crack tip at a general position s so that:

$$\sigma = K_I(s)/(2\pi r)^{1/2} = k_I(s)\, \sigma_{zz}/(2\pi r)^{1/2} \tag{1}$$

where r is the distance along the normal from the crack edge, $K_I(s)$ is the mode I opening stress intensity factor defined at general position s on the crack edge, and $k_I(s)$ is the normalized mode I stress intensity factor at s. We now define a contour integral P evaluated along the length of the crack edge c :

$$P = \int_C \rho\, k_I^2\, dl \tag{2}$$

with ρ as previously defined.

Budiansky and Rice have shown that for an elliptical crack embedded in a solid

$$k_{I_{max}} = \gamma \left(\frac{8P}{\pi^3} \right)^{1/6} \tag{3}$$

where γ = 1 for an embedded circular crack, and γ ≅ 1 to within 10% for embedded elliptical cracks with a/c > 1/20 .

For the case of elliptical shaped surface cracks, it is necessary to evaluate γ to correct for the added intensification of $k_{I_{max}}$ due to the intersection of the free surface with the crack edges. To accomplish this task, the variation in $k_I(s)$ around the crack edge must be known for different values of the aspect ratio a/c . Smith and Sorenson [3] have calculated $k_I(s)$ for elliptical shaped surface cracks subjected to a uniform tensile stress. Additionally, Smith et al. [11] have evaluated $k_I(s)$ for half penny shaped surface cracks subjected to bending stresses. Combining Eqs. (2) and (3), we obtain an expression for γ :

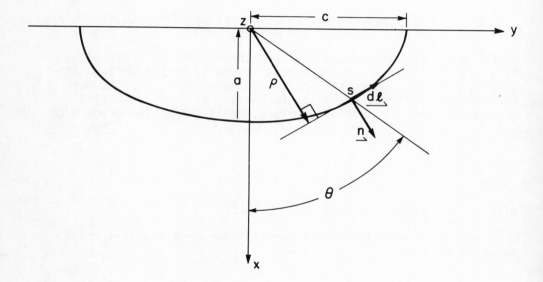

Figure 1.   Crack coordinate system.

$$\gamma = \cfrac{k_{Imax}}{\left[\cfrac{8\int_{c}\rho k_I^2 d\ell}{\pi^3}\right]^{1/6}} \qquad (4)$$

This expression may be evaluated by integrating the curves of Smith et al.[11] numerically. For a half penny shaped crack, we find $\gamma = 1.22$. This result for γ is greater than for buried cracks but is still a reasonable one to use for cracks with ratios of a/c greater than 0.80 .

We have modified the original theory of Budiansky and Rice to evaluate k_{Imax} for a surface crack by making a single measurement of the reflection coefficient of a Rayleigh wave normal to the crack plane.

The acoustic reflection coefficient of a surface crack [4] is:

$$S_{11} = \frac{j\omega}{4} \int_{A_c} \sigma_{zz}^A \Delta U_z dA \qquad (5)$$

Here ω is the frequency of the Rayleigh wave, and σ_{zz}^A is defined as the stress associated with an incident surface wave of unit power in this equation.

The simplifying assumption is made that since the crack is very small compared with the acoustic wavelength, the component of dynamic stress associated with the Rayleigh wave, σ_{zz}^A , does not diminish appreciably with increasing distance from the surface and is taken as uniform.

ΔU_z is the displacement across adjacent crack faces due to the applied stress. This integral is then evaluated over the crack area A_c . For an elliptical shaped crack, Budiansky and Rice [2] and Budiansky and O'Connell [5] have shown that

$$1/2 \int_{A_c} \Delta U_z dA = \frac{1 - \nu^2}{3E} \sigma_{zz}^A P \qquad (6)$$

where E is Young's modulus, and P has been previously defined in Eq. (2). Combining Eqs. (5) and (6), we find that

$$S_{11} = \frac{j\omega(1 - \nu^2)}{12\ E}\ \sigma_{zz}^A P \tag{7}$$

After evaluating σ_{zz}^A and P for a uniform applied stress
and $\kappa a \ll 1$, we find the following relationship between S_{11}
and a for a half penny shaped crack:

$$S_{11} = \frac{j2f_z\kappa^2 a^3}{3(1 - \nu)w} \tag{8}$$

where κ is the wave number, $2\pi/\lambda$, a is the crack radius for
a half penny shaped surface crack, f_z is a constant equal to
0.34 for glass as tabulated by Auld,[6] ν is Poisson's ratio,
and w is the width of the acoustic beam. We reiterate that this
relation is only valid for $\kappa a \ll 1$.

From this simplified scattering theory it is now possible to
derive a relation between S_{11} and the maximum value of the normalized stress intensity factor, k_{Imax} . Once k_{Imax} of a crack
is known, linear elastic fracture mechanics tells us that

$$\sigma_f = \frac{K_{Ic}}{k_{Imax}} \tag{9}$$

where σ_f is the fracture stress of the material, K_{Ic} is its
fracture toughness, and k_{Imax} is previously defined.

THE ACOUSTIC PREDICTION OF THE MAXIMUM STRESS INTENSITY FACTOR

To evaluate k_{Imax} from a Rayleigh wave scattering experiment for half penny shaped cracks and $\kappa a \ll 1$, from Eqs. (3) and
(7), we note:

$$S_{11} = \frac{j\omega(1 - \nu^2)}{12\ E}\ \sigma_{zz}^A P \tag{10}$$

Solving Eq. (10) for P and substituting into Eq. (3) yields

$$k_{Imax} = \gamma \left[\frac{96E\ |S_{11}|}{(1 - \nu^2)\ \omega\sigma_{zz}^A \pi^3} \right]^{1/6} \tag{11}$$

Finally, after evaluating σ_{zz}^A for unit power in the surface wave and correcting for diffraction loss:

$$k_{Imax} = \gamma \left[\frac{3E\lambda_R^2 z \, |S_{11}|}{(1 - \nu^2) \, \pi^5 V_s^2 w f_z [1 - (V_s/V_L)^2]^2} \right]^{1/6} \tag{12}$$

where E is Young's modulus, λ_R is the Rayleigh wavelength, z is the crack-transducer distance, and V_s and V_L are the shear and longitudinal wave velocities, and w is the acoustic beam width. We note that $k_{Imax} \propto S_{11}^{1/6}$. This implies that S_{11} need not be known very precisely because the percentage error in the measurement will be decreased by a factor of 6 when k_{Imax} is determined.

EXPERIMENTAL VERIFICATION OF THEORY

We have designed an experiment to test the basic assertions of our theory of Rayleigh wave scattering from surface cracks. The two relationships which can be evaluated experimentally are: S_{11} versus a (Eq. (8)) and S_{11} versus k_I (Eq. (12)). An experimental technique is described in which the magnitude of the reflection coefficient, S_{11}, of surface cracks is measured for cracks of different sizes. The specimens containing the acoustically measured cracks are then fractured. This allows us to measure the pre-fracture crack size from the fracture surfaces and to calculate the fracture toughness of individual specimens. We shall then show how our measured values of crack radius a, and fracture stress σ_f, compare with measured values of the crack reflection coefficient, S_{11}.

Specimen Preparation

Pyrex discs 3 mm thick by 7.6 cm in diameter were prepared for fracture toughness testing and acoustic scattering by introducing small, semi-elliptical shaped surface cracks into the center of each disc. These pre-cracks were produced by pressing a Knoop microhardness indentor into the specimen while applying a bending moment to the specimen such that the material under the indentor was in tension. This technique produced surface cracks that ranged in size from a half penny shape of 50 µm radius to semi-elliptical cracks with minor axis length of 500 µm and $a/c = 0.80$. The crack size could be varied by applying different combinations of bending, moment, and force on the Knoop indentor. These small sizes were needed to satisfy the small κa approximation in the scattering theory.

Rayleigh Wave Scattering Measurement

The experimental set-up for measuring the acoustic surface wave reflection coefficient, S_{11} , is shown schematically in Fig. 2. A wideband, high efficiency Lucite wedge transducer is used to excite and detect the surface acoustic waves. To insure reproducibility, a liquid such as water or ethylene glycol is used between the wedge and the ceramic. The transducer has a center frequency of 3.4 MHz . All measurements were taken in the far field of the transducer $(z > w^2/\lambda_R)$ where w is the acoustic beam width. Most of the acoustic measurements were taken with the front of the wedge a distance $z = 2.3$ cm from the crack and with the acoustic surface wave normal to the crack surface. In our scattering experiments, we calibrate the transducers by measuring the loss between two closely spaced transducers, L_1 and L_2 . We also measure the loss of a reflected signal from a corner near the transducer. If the reflection loss of the corner is Lc , we measure:

$$2L_1 + Lc = Lc_1 \tag{13}$$

$$2L_2 + Lc = Lc_2 \tag{14}$$

$$L_1 + L_2 = L_0 \tag{15}$$

It follows that:

$$L_1 = \frac{2L_0 - Lc_2 + Lc_1}{4} \tag{16}$$

The measured loss of an individual wedge transducer is typically 9 dBs .

Fracture Toughness Testing

The strength of each disc in biaxial flexure was determined by the method of Wachtman, et al.[7] The specimen is supported on three equally spaced balls concentric with the load, which is applied to the center of the specimen by a flat small diameter, flat end piston. The biaxial tensile surface stress at the center of the disc is calculated from an equation derived by Kirstein and Wooley, [8] and for our specimen geometry this equation reduces to:

$$\sigma_B = c \frac{L}{d^2} \tag{17}$$

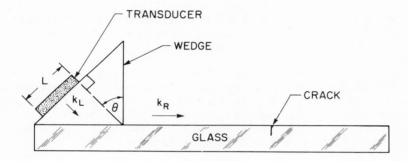

Figure 2.   Schematic diagram of surface crack and test set up.

Here, σ_B is the fracture stress, L is the load required to
fracture the disc, d is the thickness of the disc, and the constant c contains the loading radius, specimen radius, piston
radius, and Poisson's ratio ν for the material.

The discs were loaded to fracture at a loading rate of approximately 60 MPa per second using an MTS System 810 servohydraulic
testing machine operated in stroke control mode. The loading rate
was chosen to minimize possible slow crack growth effects. To
further insure against environmental influences (humidity), the
specimens were pumped down in a vacuum for one hour prior to
strength testing and subjected to a stream of dry nitrogen gas for
three minutes prior to and during the flexure test.

After the specimens were fractured, the geometry of the pre-
cracks was studied by examining the fracture surface with a metal-
lograph using reflected light at a magnification of 50X . The
aspect ratio (a/c) of the semi-elliptical surface precracks
could then be easily measured using an eyepiece with a properly
graduated reticle.

The stress intensity factor for the precrack was evaluated by
the method of Shaw and Kobayashi. [9] This analysis was chosen
because it not only takes into account the proximity of the back
surface to the precrack, but also allows for the presence of a
linearly varying stress field along the glass width. The expression for the maximum stress intensity is:

$$K_I = \frac{M_B \sigma_B (\pi a)^{1/2}}{E(k)}\tag{18}$$

where $E(k)$ is an elliptic integral of the second kind, M_B is
a magnification factor which takes into account the aspect ratio
and the proximity of the crack depth to the neutral axis, σ_B is
the maximum bending stress at the surface of the specimen, and a
is the length of the semi-minor axis (depth) of the precrack.

It should be noted that other forms for the evaluation of the
stress intensity factor were tried, but Eq. (18) yielded consistent
and conservative values of the fracture toughness which were in
good agreement with the results of Wiederhorn on Pyrex glass
fractured in a dry N_2 environment. [10]

Fracture Toughness of Pyrex Glass Determined in Biaxial Bend Tests

The 37 samples tested gave a normally distributed set of
fracture toughness measurements, and the mean value of toughness,

K_{Ic} , at a confidence level of 90% is: $K_{Ic} = 0.71 \pm 0.04$ MPa $\sqrt{m}$.
This compares favorably with Wiederhorn's [10] experimental results
of 0.75 $\pm$ 0.01 MPa $\sqrt{m}$ for 6 specimens and 0.778 $\pm$ 0.011 MPa $\sqrt{m}$
for 8 specimens. It should be noted that Wiederhorn's results were
obtained with specimens made from Pyrex glass microscope slides.

THE ACOUSTIC PREDICTION OF THE FRACTURE STRESS

We have shown previously in Eqs. (10) through (12) that the
acoustically predicted maximum normalized stress intensity factor,
k^a_{Imax} , of a surface crack can be inferred by measuring the crack
reflection coefficient, S_{11} . The acoustic prediction of the
fracture stress, σ^A_f , is related to k_{Imax} through Eq. (9). For
each specimen, an acoustic prediction of the fracture stress was
evaluated using the measuring value of S_{11a} to calculate
k_{Imax} . For all cases, k_{Imax} was corrected by a value of
$\gamma = 1.22$ for an assumed half penny shaped crack to correct for
surface stress intensification. (See Eq. (12).) This corrected
prediction of the fracture stress is compared to the mechanically
measured fracture stress for each specimen in Fig. 3. Although
some scatter is evident here due to the statistical nature of
K_{Ic} and measurement errors in S_{11} , the correlation is quite
good.

Additional insight may be obtained from the data by determining the measured normalized stress intensity factor, k^m_{Imax} ,
factor from the calculated stress intensity factor at fracture.
K_{Ic} is calculated from the measured crack size after fracture and
the measured fracture stress with

$$k^m_{Imax} = \frac{K_{Ic}}{\sigma^m_f} \qquad (19)$$

This measured value of the normalized stress intensity factor
for each crack is compared to the acoustic prediction k^a_{Imax} in
Fig. 4. We see here a trend for k^a_{Imax} to be greater than k^m_{Imax}
for virtually all specimens. This means that if K_{Ic} were equal
for all specimens, the acoustic prediction of σ^A_f would always
be conservative by roughly 20%. From an engineering point of
view this should be helpful in determining a safety factor for
conservative fracture prevention.

SOME OBSERVATIONS OF S_{11} VERSUS k_{Imax}

To understand conceptually the reason for the predicted value
of k^a_{Imax} to be higher than that of k^m_{Imax} , it is enlightening
to observe graphically a comparison of S_{11} with k^m_{Imax} . In

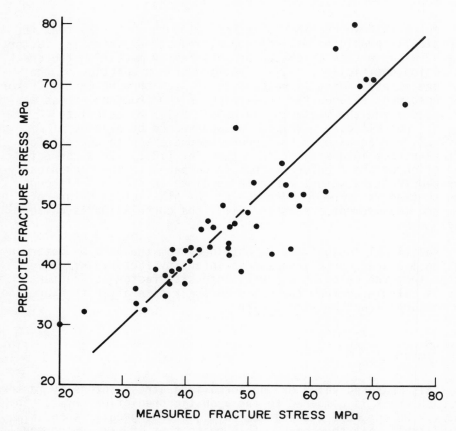

Figure 3. Comparison of acoustically predicted fracture stress
σ_f^A with measured value of fracture stress σ_f^m for
each specimen.

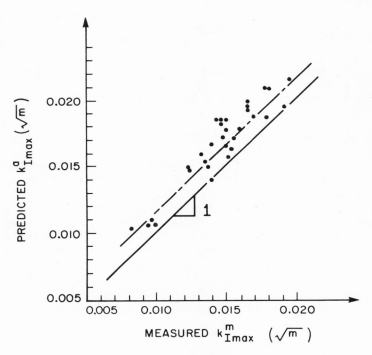

Figure 4. Comparison of the acoustically predicted maximum
normalized stress intensity factor, k^a_{Imax}, with
the measured value k^m_{Imax}.

Fig. 5, we plot S_{11} versus $\log k_{Imax}^m$ to show that the slope of
the regression line of the experimental data is very close to the
theoretically predicted slope of 6. This is a surprising result,
since $S_{11} \sim k_I^6$ is only theoretically valid for scattering conditions with $\kappa a << 1$, and the data in Fig. 5 represents cracks
with $0.3 \leq \kappa a \leq 3$. The difference in y intercept between the
theory and regression lines is caused by the multiplication of
the S_{11} versus k_{Imax} relationship for a volume crack by a surface stress magnification factor. The value of γ used was calculated for half penny shaped surface cracks in a finite solid
subjected to bending stresses. In all cases, this calculated
stress magnification factor was larger than the measured stress
magnification factor evaluated from the post-fracture measurement
of the crack dimensions. At a given value of S_{11} , the theory
predicts that the crack will be slightly more intense than will be
actually observed. This empirical result is very important in a
predictive sense for it expands the range of crack sizes it is
possible to measure with this surface wave technique.

THE USE OF THE REFLECTION COEFFICIENT TO DETERMINE CRACK SIZE

        In Eq. (8) we established a relationship between the reflec-
tion coefficient S_{11}, and the half penny shaped crack radius of
a , such that $S_{11} \propto a^3$. In our simplified scattering theory,
the magnitude of S_{11} is determined by an acoustically perceived
half penny shaped crack. In reality, the cracks which we used
had crack aspect ratios of $0.80 \leq a/c \leq 1.2$. Since the magnitude of S_{11} tells us nothing about a/c , we need to find
another kind of size variable to relate to S_{11} . For a bulk wave
incident to an elliptical crack completely embedded in a solid,

$$S_{11} \propto P = \frac{4\pi ca^2}{E(k)} \tag{20}$$

        We define a half penny shaped crack with radius a' (an
effective half penny shaped crack radius) such that P_E for a
half ellipse with a particular a/c value would be the same as
P_{HP} for a crack with $a' = a = c$. Then, from the Budiansky and
Rice theory for a buried crack,

$$P_E = P_{HP} = \frac{ca^2}{E(k)} = \frac{2a'^3}{\pi} \tag{21}$$

Finally, the effective radius of the crack can be estimated to be
a' , where

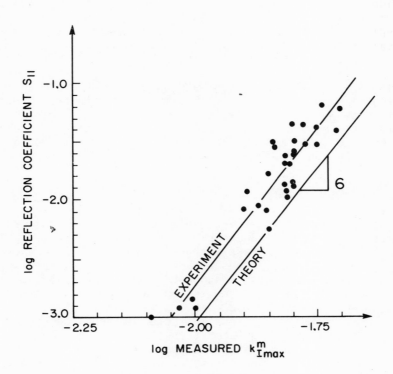

Figure 5.   Log of scattering coefficient S_{11} versus log of measured k^m_{Imax} .

$$a' = \left[\frac{\pi c a^2}{2E(k)} \right]^{1/3} \tag{22}$$

where $E(k)$ has been previously defined.

Now substitution into Eq.(8) produces a theoretical relationship between a' and S_{11} such that $S_{11} \propto a'^3$. In Fig. 6 we have plotted $\log S_{11}$ versus $\log a'$. We see that the theoretical prediction of S_{11} versus a' intersects the regression line of the experimental data at a small κa value and that the experimental values of a' become significantly less than their predicted values at large κa .

This is indicative of the breakdown of the theoretical assumption of constant acoustic stress varying with crack depth, inherent in small κa acoustic theory. It is encouraging that for small κa measurements the average error in size prediction is less than 20%. To measure larger cracks accurately would clearly require longer wavelength acoustic probing.

CONCLUSIONS

In the present work we have outlined an extension of the bulk wave scattering theory of Budiansky and Rice [2] which allows us to predict the size and stress intensity factor of surface cracks with reasonable accuracy. We stress that this theory is designed to be most accurate for scattering conditions such that $\kappa a \ll 1$. We have obtained reasonable results for surface cracks in Pyrex glass specimens for which $0.3 \leq \kappa a \leq 3$. In all cases, the maximum normalized stress intensity factor as predicted by our acoustic measurements was greater than the measured value of maximum normalized stress intensity factor. Additionally, the prediction of crack size from measurements of S_{11} is seen to be accurate for the small κa range only.

It is to be expected that S_{11} versus a would be less accurate than S_{11} versus k_I at large κa values since $a \propto k_I^2$. Nevertheless, we have shown that the size of surface cracks can be predicted with reasonable accuracy when scattering conditions are maintained as specified in the long wavelength theory.

Finally, from our correlation between the predicted and measured value of the fracture stress, the average error between experiment and theory is found to be less than 20%. This result gives convincing evidence that Rayleigh wave scattering from surface flaws is an effective NDE technique and therefore deserves further study in other ceramic materials.

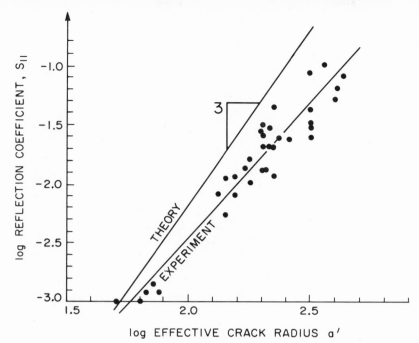

Figure 6.   Log of scattering coefficient S_{11} versus log of effective crack radius a' .

REFERENCES

1. J. Tien, B. Khuri-Yakub, and G. S. Kino, Proc. ARPA/AFML Rev. of Progress in Quantitative NDE, La Jolla, Calif., July 1979.
2. B. Budiansky and J. Rice, J. Appl. Mech., 45(78).
3. F. W. Smith and D. R. Sorenson, Int. J. of Fracture, 12(76), 47-57.
4. G. S. Kino, J. Appl. Phys., 49(78) 3190-3199.
5. B. Budiansky and R. J. O'Connell, Int. J. Solid Structures, 12(76), 81-97.
6. B. A. Auld, Acoustic Fields and Waves in Solids, vols. I & II (Wiley-Interscience, New York, 1973).
7. J. B. Wachtman, Jr., W. Capps, and J. Mandel, J. Mater. Sci., 7(72), 188-194.

8. A. F. Kirstein and R. M. Wooley, J. Res. NBS, C71(67), 1-10.
9. R. C. Shaw and A. S. Kobayashi, ASTM STP 513 (Amer. Soc. for Testing Materials, Philadelphia, Penn., 1972), pp. 18-19.
10. S. M. Wiederhorn, J. Am. Ceram. Soc., 52(69), 99-105.
11. F. W. Smith, A. F. Emery, and A. S. Kobayashi, JAM (67), 953-959.

ANISOTROPIC ELASTIC CONSTANTS OF A FIBER-REINFORCED

BORON-ALUMINUM COMPOSITE

S. K. Datta

Department of Mechanical Engineering, University of
Colorado, Boulder, Colorado, 80309, USA

H. M. Ledbetter

Fracture and Deformation Division
National Bureau of Standards
Boulder, Colorado, 80303, USA

Abstract

Elastic constants, both the C_{ij}'s and the S_{ij}'s, were measured
and calculated for a laminated, uniaxially fiber-reinforced
boron-aluminum composite. Three theoretical models were considered:
square-array, hexagonal-array, and random-distribution. By
combining several existing theoretical studies on randomly distributed fibers, a full set of elastic constants can be predicted
for this model. The random-distribution model agrees best with
observation, especially for off-diagonal elastic constants.
Considering all nine elastic constants, observation and theory
differ on the average by six percent.

1. Introduction

Fiber-reinforced composites have been studied extensively,
both experimentally and theoretically. Recent reviews on this
subject include that of experiment by Bert[1] and of theory by
Sendeckyj[2].

Two types of fiber-reinforced composites occur in practice:
continuous-fiber and short (chopped)-fiber. The former type has
been studied more thoroughly and constitutes the object of this
study.

Boron-fiber-reinforced aluminum, a so-called advanced-technology composite, was the particular material that was studied. The principal purpose of the study was to evaluate various models for predicting the composite's macroscopic elastic constants from those of its constituents. The composite's elastic constants were measured two ways: by ultrasonic-velocity measurements, which yield the C_{ij}'s, the elastic stiffnesses; and by standing-wave resonance, which yields the S_{ij}'s, the elastic compliances. These macroscopic elastic constants define for most cases the mechanical response of the composite to an applied force.

2. Material

The studied composite consisted of 0.14-mm-diameter boron fibers in an aluminum-alloy-6061 matrix. The alloy was in the F-tempered condition, as diffusion bonded. The composite, containing 48 percent fibers by volume, was fabricated as a 10 x 10 x 1.1 cm plate containing about seventy plies. Figure 1 is a photomicrograph showing the distribution of fibers in the transverse plane. Cracks in the boron fibers developed during specimen preparation and do not exist in the bulk material. Mass density, measured hydrostatically, was 2.534 g/cm^3.

Throughout this study we adopt the following co-ordinate system: x_3 is the fiber direction, x_1 is normal to the laminae plane, and x_2 is orthogonal to x_1 and x_3. Thus, x_1 and x_2 are equivalent directions from the viewpoint of all three theoretical models considered here.

3. Experimental Methods

Our experimental methods were chosen to provide the advantages of small specimens and low inaccuracy.

For brevity, experimental details are omitted here; they were described previously for both the ultrasonic-velocity method by Ledbetter, Frederick, and Austin[3] and the resonance method by Ledbetter[4].

Briefly, the ultrasonic-velocity method consisted of a pulse-echo technique using gold-plated piezoelectric crystals at frequencies near 10 MHz with specimen thickness varying from 0.2 to 1.0 cm. Except for an improved velocity-measurement system used in this study, it proceeded similarly to that described by Ledbetter and Read[5].

The resonance method used a three-component (Marx) oscillator. Specimens were rod-shaped, about 0.4 cm in either circular or square cross section, and 4 to 10 cm long. Frequencies ranged from 30 to 50 kHz.

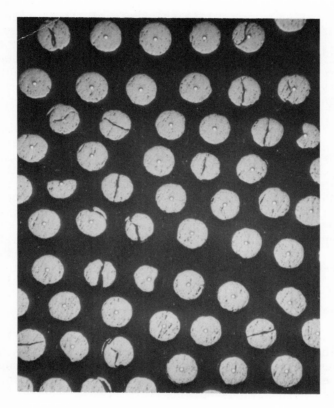

Fig. 1.   Photomicrograph showing distribution of 0.14-mm diameter
          boron fibers in an aluminum alloy matrix. Plane of photo
          is perpendicular to fibers.

4. Experimental Results

Results of the experimental studies are given in Table 1 and in Table 2 for the C_{ij}'s and the S_{ij}'s, respectively. The nine C_{ij}'s are based on a least-squares fit to eighteen separate wave-velocity propagation directions and polarizations. For details, see Ledbetter and Read[5]. The five S_{ij}'s are based on extensional-mode and torsional-mode measurements on cylindrical specimens whose axes were in the x_2-x_3 plane, as described by Read and Ledbetter[6]. The plate thickness of the studied material was too short to permit S_{ij} measurements along x_1. Thus, the measured S_{ij}'s are incomplete since they do not reflect the full orthotropic symmetry of the composite. However, as shown in Table 2, except for S_{66} and S_{12}, this incomplete set of S_{ij}'s differs only slightly from those derived from the observed C_{ij}'s, which do contain the full orthotropic symmetry.

Table 3 shows the elastic constants of the constituent materials, both being assumed isotropic. In this Table, ρ denotes mass density, E Young's modulus, μ shear modulus, B bulk modulus (reciprocal compressibility), k plane-strain bulk modulus for lateral dilatation without longitudinal extension, and λ the Lamé constant. For aluminum these parameters were determined ultrasonically, as described above, on a bulk specimen; their uncertainty is less than one percent. Boron's elastic constants are less certain, a wide range of values being reported in the literature. The boron values in Table 3 arose from fitting a linear rule-of-mixtures to our observed S_{33} value combined with Gschneidner's[7] recommended values for boron's bulk modulus.

5. Square-array Model

For a periodic square-lattice array of long, parallel, circular fibers, using a homogeneous continuum model, Achenbach[8] derived three relationships for plane, harmonic wave-propagation velocities expressed as ρv^2. Simplified and modified for present purposes, and with obvious notational correspondences, these are as follows.

For a longitudinal wave propagating along x_3, parallel to the fibers,

$$\rho v^2 = C_{33} = a_1 - \frac{a_3^2}{a_2}, \tag{1}$$

where

$$a_1 = c(\lambda_f + 2\mu_f) + (1 - c)(\lambda_m + 2\mu_m),$$

$$a_2 = 4c(\lambda_f + \mu_f), \text{ and}$$

$$a_3 = 2c\lambda_f.$$

For a longitudinal wave propagating along x_1, perpendicular to the fibers,

$$\rho v^2 = C_{11} = C_{22} = \mu_m/4M, \tag{2}$$

where

$$M = \frac{c[cD - (1 - c)\,B] + (1-c)\,[(1 - c)\,A - cB]}{4\,(AD-B^2)},$$

$$A = a_{22_f} - \frac{(1 - c)^2}{c}\,a_{23_f}^2\,\frac{d}{q}\,\frac{1}{\mu_m},$$

$$B = (1-c)\,a_{23_f}\,a_{23_m}\,\frac{d}{q}\,\frac{1}{\mu_m},$$

$$D = \left[a_{22_m} - c\,a_{23_m}^2\,\frac{d}{q}\right]\frac{1}{\mu_m},$$

$$\frac{d}{q} = \frac{c}{(1-c)^2\,a_{22_f} + c^2\,a_{22_m}},$$

$$a_{22_f} = c\,(\lambda_f + 2\mu_f),$$

$$a_{23_f} = c\lambda_f,$$

$$a_{22_m} = (1-c)\,(\lambda_m + 2\mu_m) + \frac{2}{\pi(4-\pi)}\,c\,\mu_m + \left[(\frac{\pi}{c})^{\frac{1}{2}} - 2\right]\frac{c}{\pi}\,\mu_m,$$

and

$$a_{23_m} = (1-c)\,\lambda_m + \frac{4}{\pi(4-\pi)}\,c\,\mu_m.$$

For a transverse wave propagating along x_3, parallel to the fibers, and polarized along x_1, perpendicular to the fibers,

$$\rho v^2 = C_{55} = C_{44} = a_1 - \frac{a_2^2}{a_3} \tag{3}$$

where

$$a_1 = c\,\mu_f + (1-c)\,\mu_m,$$

$$a_2 = c\,(\mu_f - \mu_m),\ \text{and}$$

$$a_3 = c\,\mu_f + \frac{c^2}{1-c}\,\mu_m$$

The Lamé constants, λ and μ, of the fiber and matrix are denoted by subscripts f and m; c denotes the volume fraction of
fibers, 0.48 for the present case, and equals $\pi r^2/d^2$, where r denotes fiber radius and d distance between fiber centers in a square
lattice. Clearly, r/d = .3909, or d/r = 2.5583. The constants
a_1, a_2, a_3, etc. arise from the strain-energy density in Achenbach's
derivation.

Results predicted using these relationships based on
Achenbach's square-array model are shown in Table 1.

6. Hexagonal-array Model

In calculations similar to Achenbach's, the five independent
elastic constants for a composite consisting of parallel fibers
arranged hexagonally were derived by Hlavacek[9]. His relationships,
rearranged for present purposes, are:

For a longitudinal wave propagating along x_1, perpendicular
to the fibers,

$$\rho v^2 = C_{11} = C_{22} = a_1 - \frac{2a_7 a_8 a_{23} - a_{22}\,(a_7^2 + a_8^2)}{a_{22}^2 - a_{23}^2}, \tag{4}$$

where

$$a_1 = (3V + 1)\,\lambda_m + (7V + 2)\,\mu_m,$$

$$a_7 = -V\,(3\lambda_m + 7\mu_m),$$

$$a_8 = -V\,(\lambda_m + \mu_m),$$

$$a_{22} = -c \ (\lambda_f + 2\mu_f) - (3V - c) \ \lambda_m - (7V - 2c) \ \mu_m,$$

$$a_{23} = -c \ \lambda_f - (V-c) \ \lambda_m - V\mu_m, \ \text{and}$$

$$V = \frac{-c}{8(1 - c^{1/2})^2} \ \ln c.$$

For a longitudinal wave propagating along x_3, parallel to the fibers,

$$\rho v^2 = C_{33} = a_{15} + \frac{2a_{18}^2}{a_{22} + a_{23}}, \tag{5}$$

where

$$a_{15} = c(\lambda_f + 2\mu_f) + (1-c) \ (\lambda_m + 2\mu_m) \ \text{and}$$

$$a_{18} = c \ (\lambda_f - \lambda_m).$$

For a transverse wave propagating along x_1, perpendicular to the fibers, and polarized along x_3, parallel to the fibers,

$$\rho v^2 = C_{44} = C_{55} = a_3 + \frac{a_{10}^2}{a_{28}}, \tag{6}$$

where

$$a_3 = c\mu_f + (1-c) \ \mu_m,$$

$$a_{10} = c(\mu_f - \mu_m), \ \text{and}$$

$$a_{28} = -c\mu_f - (4V-c) \ \mu_m.$$

For a transverse wave propagating along x_1, perpendicular to the fibers, and polarized along x_2, perpendicular to the fibers,

$$\rho v^2 = C_{66} = -a_2 + \frac{2a_8 \ a_9 \ a_{26} - a_{25} \ (a_8^2 + a_9^2)}{a_{25}^2 - a_{26}^2}, \tag{7}$$

where

$$a_2 = V\lambda_m + (5V + 1) \ \mu_m,$$

$$a_9 = -V(\lambda_m + 5\mu_m),$$

$$a_{25} = -c\mu_f - V\lambda_m - (5V-c)\,\mu_m, \text{ and}$$

$$a_{26} = -c\mu_f - V\lambda_m - (V-c)\,\mu_m.$$

For the fifth independent elastic constant, Hlavacek derived an expression for $C_{13} + C_{44}$, which is related to the propagation of a transverse wave:

$$C_{13} = \frac{a_5\,a_{28} + a_{10}\,a_{19}}{a_{28}} + \frac{a_{18}\,(a_7 + a_8)}{(a_{22} + a_{23})} - C_{44}, \tag{8}$$

where

$$a_5 = \lambda_m + \mu_m$$

and

$$a_{19} = -4V\mu_m.$$

In Hlavacek's model, the fiber volume concentration, c, equals r_1/r_2, the ratio of the fiber radius to that of the circle radius that equals the hexagonal "unit-cell" area. Constants a_1, a_2, a_3, etc. arise from Hlavacek's equations of motion.

Results from Hlavacek's hexagonal-array model are shown in Table 1.

7. Random-distribution Model

Considering the propagation of time-harmonic elastic waves in a composite of circular cross section, parallel fibers distributed randomly in a matrix, Bose and Mal[10,11] derived the following expressions for three elastic moduli.

For a longitudinal wave propagating along x_1, perpendicular to the fibers,

$$\rho v^2 = C_{11} = C_{22} = \frac{[1 - c\,(1-c^2)\,P_2 - 2c^2\,P_oP_2]\,(\lambda_m + 2\mu_m)}{(1 + cP_o)\,[1 + c\,(1 + c^2)\,P_2]}, \tag{9}$$

where $P_o = \dfrac{\lambda_m + \mu_m - \lambda_f - \mu_f}{\lambda_f + \mu_f + \mu_m}$ and

$$P_2 = \frac{\mu_m (\mu_m - \mu_f)}{\mu_f (\lambda_m + 3\mu_m) + \mu_m (\lambda_m + \mu_m)} .$$

For a transverse wave propagating along x_1, perpendicular to the fibers, and polarized along x_3, parallel to the fibers,

$$\rho v^2 = C_{44} = C_{55} = \left[1 + \frac{2c (\mu_f - \mu_m)}{\mu_f + \mu_m - c(\mu_f - \mu_m)}\right]\mu_m . \qquad (10)$$

For a transverse wave propagating along x_2, perpendicular to the fibers, and polarized along x_1, perpendicular to the fibers,

$$\rho v^2 = C_{66} = \frac{1}{2}(C_{11} - C_{12})$$

$$= \left[1 + \frac{2c(\lambda_m + 2\mu_m)(\mu_f - \mu_m)}{2\mu_m(\lambda_m + 2\mu_m) + (1-c)(\lambda_m + 3\mu_m)(\mu_f - \mu_m)}\right]\mu_m \qquad (11)$$

Results from the Bose-Mal random-distribution model are shown in Table 1.

8. Full Random-distribution Model

In this section we show that combining the Bose-Mal relationships with previously derived relationships for other elastic constants leads to a full set of five independent elastic constants for the random-distribution case. For this, a new notation (see Sendeckyj[2]) is useful, which will be described as it is introduced.

Hashin and Rosen[12] derived relationships for the effective moduli of a continuous-fiber-reinforced composite where the fibers are distributed randomly and homogeneously. Hashin[13] gave these relationships in essentially the forms

$$k_T = k_m - \frac{c(k_f - k_m) (k_m + \mu_m)}{k_f + \mu_m - c(k_f - k_m)} \qquad (12)$$

and

$$\mu_{LT} = \mu_m + \frac{2c(\mu_f - \mu_m)\mu_m}{\mu_f + \mu_m - c(\mu_f - \mu_m)} . \qquad (13)$$

Here, k denotes the two-dimensional, plane-strain bulk modulus, which is $\lambda + \mu$ in an isotropic material. Subscript T denotes deformation in the x_1-x_2 plane, perpendicular to the fibers. Subscript LT denotes shear deformation in any plane containing the x_3 axis, that is, the fiber axis. Equations (12) and (13) were also derived by Hill[14] for a single fiber placed concentrically in a cylindrical matrix. Hermans[15] obtained them also by considering Hill's assembly to be embedded in an unbounded solid having the effective elastic moduli of the composite. In this model, the fiber:matrix-cylinder radius ratio is $r_1/r_2 = c^{1/2}$. In the limit $r_1 = 0$, Hermans' model becomes the "self-consistent" model of Hill (1965).

Knowing k_T, one can derive the Young's modulus E_L (along x_3) and the Poisson's ratio $\nu_{LT} = -\varepsilon_1/\varepsilon_3$, where uniaxial stress is along x_3 and ε denotes strain. This derivation requires some exact relationships given by Hill[14]:

$$\frac{\frac{1}{k_T} - \frac{1}{k_f}}{\nu_{LT} - \nu_f} = \frac{\frac{1}{k_T} - \frac{1}{k_m}}{\nu_{LT} - \nu_m} = \frac{\frac{1}{k_f} - \frac{1}{k_m}}{\nu_f - \nu_m} = \frac{4}{E_L} \frac{\nu_{LT} - \nu_m - c(\nu_f - \nu_m)}{E_L - E_m - c(E_f - E_m)} \qquad (14)$$

Solving for E_L and V_{LT}, using Eq. (12), one finds

$$E_L = (1-c)E_m + cE_f + \frac{4c (1-c)(\nu_f - \nu_m)^2}{\frac{1-c}{k_f} + \frac{c}{k_m} + \frac{1}{\mu_m}} \qquad (15)$$

and

$$\nu_{LT} = (1-c)\nu_m + c\nu_f + \frac{c(1-c)(\nu_f - \nu_m)(\frac{1}{k_m} - \frac{1}{k_f})}{\frac{(1-c)}{k_f} + \frac{c}{k_m} + \frac{1}{\mu_m}} \qquad (16)$$

Of course, both of these relationships represent a simple, linear rule- of- mixtures plus a higher-order correction term. The advantages of Eqs. (12)-(13) are clear: they give simple, explicit relationships for four of the composite's elastic constants in terms of the isotropic elastic constants of the constituent materials, the fiber and the matrix. On the other hand, self-consistent approaches lead to implicit relationships.

Bose and Mal[10,11] also obtained Eqs. (12) and (13), but by a different approach: by considering the effective velocity for long-wavelength plane waves propagating perpendicular to the fibers, for both a longitudinal wave and for a shear wave polarized along x_3, the fiber axis. By a similar approach, examining a shear wave propagating and polarized in a plane transverse to the fibers, Bose and Mal[10] also derived an expression for the shear modulus μ_{TT}:

$$\mu_{TT} = \mu_m + \cfrac{2c(k_m + \mu_m)(\mu_f - \mu_m)}{k_m + (k_m + 2)[c\mu_m + (1-c)\mu_f]}{\mu_m} \qquad (17)$$

Previous studies (Hashin and Rosen,[12] and Hill[14]) did not obtain this explicit expression, but bounds instead. In fact, Eq. (17) corresponds to their lower bound. Equations (12), (13), (15), and (16) are also identical to the lower-bound results. Thus, the quasi-crystalline approximation used by Bose and Mal[10,11] to derive Eqs. (12), (13), and (16) leads to the lower-bound results, which these authors believe to be "nearer to the actual value if correlations are ignored." Equation (17) coincides also with Hermans'[15] result, which he points out "lies between the bounds derived by Hashin and Rosen." From the above relationships and results in Table 3 one obtains for the boron-aluminum composite:

$$E_L = 2.292 \cdot 10^{11} \text{ N/m}^2,$$

$$k_T = 1.267 \cdot 10^{11} \text{ N/m}^2,$$

$$\mu_{LT} = 0.559 \cdot 10^{11} \text{ N/m}^2,$$

$$\mu_{TT} = 0.523 \cdot 10^{11} \text{ N/m}^2, \text{ and}$$

$$\nu_{LT} = 0.230$$

If the composite behaves transversely isotropic with the unique symmetry axis along x_3, then the C_{ij} elastic-stiffness constants can be computed from

$$C_{11} = C_{22} = k_T + \mu_{TT}, \qquad (18)$$

$$C_{33} = E_L + 2\nu_{LT} C_{13}, \qquad (19)$$

$$C_{44} = C_{55} = \mu_{LT}, \tag{20}$$

$$C_{66} = \frac{1}{2}(C_{11} - C_{12}) = \mu_{TT}, \tag{21}$$

$$C_{12} = k_T - \mu_{TT}, \text{ and} \tag{22}$$

$$C_{13} = C_{23} = 2\nu_{LT} k_T. \tag{23}$$

These results are given in the final column of Table 1.

9. Discussion

Here we consider principally the correspondence between observed C_{ij}'s and their counterparts predicted by three models: square-array, hexagonal-array, and random-distribution.

C_{33} is predicted reasonably well by all three models, although all are slightly high, perhaps reflecting the slight uncertainty in Young's modulus of the boron fibers.

C_{11} is predicted best by the lattice models, and C_{22} is predicted best by the random model. As shown in Fig. 1, this probably reflects the different imperfections in fiber distributions along the x_1 and x_2 directions.

C_{44} and C_{55}, which represent shears with the force parallel to the fibers, are not predicted accurately by the square-array model; but they are predicted by both the hexagonal-array and the random models. Thus, these elastic constants permit one to distinguish fiber arrays that are transversely anisotropic (tetragonal, orthorhombic) from those that are transversely isotropic (hexagonal, random).

C_{66}, which represents shears with a force perpendicular to the fibers, is predicted well only by the random model. Since C_{66} represents also the velocity of a plane wave both propagated and polarized in the transverse plane, of all elastic constants, it should be most sensitive to the fiber distribution. Thus, the different predictions for the hexagonal and random models, both of which are transversely isotropic, are not too surprising. Furthermore, Fig. 1 shows little evidence of either three-fold or six-fold symmetry around the fiber axis.

C_{12}, C_{13}, and C_{23}, the so-called off-diagonal elastic constants are predicted surprisingly well by the random model. Since these constants are always determined indirectly, whether experimentally or theoretically, one expects larger errors than for the diagonal elastic constants, the C_{ii}'s, discussed above. Absence of large discrepancies between observation and theory for these three elastic constants gives one confidences in both the experimental measurements and in the random-distribution model for this particular composite.

Finally, note that similar observations hold also for the S_{ij}'s, shown in Table 2. However, these comparisons are less useful because only five S_{ij}'s were determined experimentally, compared with nine C_{ij}'s. Also, the experimental S_{66} is inconsistent with that predicted or that computed by inverting the C_{ij} matrix; this reflects also in the experimental S_{12} value and is believed to be an experimental artifact, related perhaps to impure torsional modes about the x_2 axis. Except for this discrepancy, there are no differences between the 30-kHz S_{ij} results and the 10-MHz C_{ij} results; thus, dispersion does not occur in this material in this frequency range. This absence of dispersion is especially interesting for the higher frequencies where the wavelength is not much larger than the fiber spacing or the fiber diameter.

ACKNOWLEDGMENT

This study was supported by the NBS Office of Nondestructive Evaluation. Most of the sound-velocity measurements were made by M. W. Austin. A first draft of the manuscript was completed while H. M. L. visited the Institute for Theoretical and Applied Physics at the University of Stuttgart. J. Dahnke contributed valuably to constructing the manuscript.

Table 1. Observed and predicted C_{ij} elastic stiffnesses of a boron-aluminum unidirectionally fiber-reinforced composite, in units of 10^{11} N/m^2.

ij	Observed	Square Model[a]	Hexagonal Model[b]	Random Model[c]	Full Random Model
11	1.852	1.856	1.872	1.790	1.790
22	1.797	1.856	1.872	1.790	1.790
33	2.450	2.480	2.551	-	2.560
44	0.566	0.451	0.561	0.559	0.559
55	0.569	0.451	0.561	0.559	0.559
66	0.526	-	0.606	0.523	0.523
12	0.779	-	0.661	-	0.745
13	0.606	-	0.578	-	0.583
23	0.604	-	0.578	-	0.583

[a]After Achenbach [8].

[b]After Hlavacek [9].

[c]After Bose and Mal [11].

Table 2. Observed and predicted S_{ij} elastic compliances of a boron-aluminum unidirectionally fiber-reinforced composite, in units of 10^{-11} m^2/N.

ij	Observed[a]	From C_{ij}'s[b]	Predicted, Full Random Model
11	0.708	0.684	0.699
22	0.708	0.707	0.699
33	0.438	0.461	0.436
44	1.755	1.767	1.788
55	1.755	1.757	1.788
66	2.123	1.901	1.913
12	-0.353	-0.261	-0.258
13	-0.106	-0.105	-0.100
23	-0.106	-0.110	-0.100

[a]Assumed transverse-isotropic symmetry.

[b]Based on orthotropic symmetry.

Table 3. Properties of constituent materials.

	Boron	Aluminum
$\rho(g/cm^3)$	2.352	2.702
$E(10^{11}N/m^2)$	3.979	0.715
$\mu(10^{11}N/m^2)$	1.763	0.267
$B(10^{11}N/m^2)$	1.785	0.751
ν	0.129	0.341
$k(10^{11}N/m^2)$	2.383	0.840
$\lambda(10^{11}N/m^2)$	0.590	0.573

References

1. C. W. Bert, Experimental characterization of composites, in: "Composite Materials, Volume 8, Structural Design and Analysis", Academic, New York (1975).
2. G. P. Sendeckyj, Elastic behavior of composites, in: "Mechanics of Composite Materials, Volume 2", Academic, New York (1974).
3. H. M. Ledbetter, N. V. Frederick, and M. W. Austin, Elastic-constant variability in stainless-steel 304, J. Appl. Phys. 51:305(1980).
4. H. M. Ledbetter, Dynamic elastic modulus and internal friction in fibrous composites, in: "Nonmetallic Materials and Composites at Low Temperatures", Plenum, New York (1979).
5. H. M. Ledbetter and D. T. Read, Orthorhombic elastic constants of an NbTi/Cu composite superconductor, J. Appl. Phys. 48:1874 (1977).
6. D. T. Read and H. M. Ledbetter, Elastic properties of a boron-aluminum composite at low temperatures, J. Appl. Phys. 48:2827 (1977).
7. K. A. Gschneidner, Physical properties and interrelationships of metallic and semimetallic elements, in: "Solid State Physics, Volume 16", Academic, New York (1964).
8. J. D. Achenbach, Generalized continuum models for directionally reinforced solids, Arch. Mech. 28:257 (1976).
9. M. Hlavacek, A continuum theory for fibre-reinforced composites, Int. J. Solids Structures 11:199 (1975).

10. S. K. Bose and A. K. Mal, Longitudinal shear waves in a fiber-
       reinforced composite, Int. J. Solids Structures 9:1075
       (1973).
11. S. K. Bose and A. K. Mal, Elastic waves in a fiber-reinforced
       composite, J. Mech. Phys. Solids 22:217 (1974).
12. Z. Hashin and B. W. Rosen, The elastic moduli of fiber-
       reinforced materials, J. Appl. Mech. 31:223 (1964).
13. Z. Hashin, Theory of composite materials, in: "Mechanics of
       Composite Materials", Pergamon, Oxford (1970).
14. R. Hill, Theory of mechanical properties of fibre-
       strengthened materials - III. Self-consistent model,
       J. Mech. Phys. Solids: 189 (1965).
15. J. J. Hermans, The elastic properties of fiber reinforced
       materials when the fibers are aligned,
       Proc. K. Ned. Akad. Wet. B70:1 (1967).

NONDESTRUCTIVE EVALUATION OF THE EFFECTS OF DYNAMIC

STRESS PRODUCED BY HIGH-POWER ULTRASOUND IN MATERIALS

Richard B. Mignogna and Robert E. Green, Jr.

Materials Science Department
The Johns Hopkins University
Baltimore, Maryland   21218

ABSTRACT

Results of measurements made during high-power ultrasonic insonation of metal specimens give new insight into the mechanical and thermal processes involved. Changes observed in the ultrasonic attenuation of test specimens during insonation indicate that high-power ultrasound can create and/or move dislocations. Measurements of the ultrasonic velocity indicate change in the moduli of the material during insonation. Heating of the test specimen to various degrees, depending upon both the specimen configuration and material has also been observed during high-power ultrasonic insonation. Details of these and further results are discussed.

INTRODUCTION

Various effects of high-power ultrasound on materials have been observed and used for many years. Some of these are plastic welding, metal welding, and metal fatigue. One of the most intriguing effects is the apparent "softening" of metals during simultaneous deformation and high-power insonation. This "softening" of metals was first reported by Blaha and Langenecker[1] in 1955, who observed an immediate reduction in the tensile load necessary to cause further deformation of zinc single crystals during the application of high-power ultrasound. Upon termination of the ultrasound the tensile load returned to the value it has prior to insonation. These observations initiated a great deal of new research into the

231

effects of large amplitude, high frequency vibrations on
the plastic deformation of metals. Since that time many
other investigators have made similar observations during
simultaneous deformation and high-power ultrasonic insonation of various metals. Much controversy pertaining to
the mechanism causing the reported load decrease or
"softening" of metals undergoing deformation and simultaneous application of high-power ultrasonic vibrations
also developed in the years following Blaha and Langenecker's[1] initial publication. The interest in this
apparent softening effect stems from both the possible
basic inferences into the nature of plastic deformation
and the potential applications to many industrial processes involving plastic deformation.

Three mechanisms which have been proposed to account
for the observed volume effect are simple superposition
of the quasistatic stress and the dynamic stress produced
by the high-power ultrasound, heating of the specimen
caused by the applied high-power ultrasound, and interaction of the high-power ultrasonic waves with dislocations in the test specimens. Although the proposed
mechanisms causing this "softening" of metals are still
somewhat controversial, high-power ultrasound is currently
being used in tube drawing, wire drawing, and other
related industrial processes.

In deformational processes involving dies or forming
tools, the softening effect may be broadly divided into
two categories: volume effect and surface effect[2]. The
influence of high-power ultrasound on the mechanisms of
deformation are included in the volume effect. The surface effect includes possible interactions of the high-
power ultrasonic vibrations with frictional forces between the specimen or work piece and the forming tool or
die.

A series of experiments have been conducted in order
to determine the actual mechanisms causing the volume
softening effect. A sufficient number of parameters to
test all of the proposed mechanisms for the influence of
high-power ultrasound on metal specimens were simultaneously measured or controlled. The parameters either
measured or controlled were the test specimen temperature,
low-power ultrasonic (8 MHz) wave velocity (directly
related to elastic moduli) and attenuation (directly
related to dislocation motion), the applied tensile load,
specimen elongation, contact pressure between the high-
power ultrasonic horn (20 kHz) and test specimen, electrical power supplied to the horn, and the high-power

insonation time.

EXPERIMENTAL SYSTEM

Figure 1 shows a block diagram of the experimental system[3]. A more detailed schematic diagram of the ultrasonic horn-converter assembly, specimen, gripping system and measuring devices is presented in Fig. 2. During part of the series of experiments the upper end of the test specimen (L of Fig. 2) was gripped and coupled to the load cell of an Instron Tensile Testing Machine[a] by a steel cage (C). A second, larger steel cage (B), bolted to the moving crosshead of the Instron (A), gripped the lower end of the specimen and also served as a mount for the high-power ultrasonic horn-converter assembly (I, J, and K). The horn-converter assembly was mounted in such a manner that it had freedom of motion along the tensile axis within the confines of the bottom grip cage. Positioning of the horn-converter assembly was accomplished with two steel guide tubes (E) and a screw (F) threaded into the bottom plate (G) of the lower grip cage and passed through the Instron crosshead (A). The horn-converter assembly was positioned so that the tip of the catenoidal horn (K) was in contact with the bottom end of the specimen. On the lower side of the crosshead a pulley (H) fastened to the horn-converter positioning screw (F), combined with a system of additional pulleys and a weight pan (not shown), permitted a constant torque to be applied to the screw. The screw, in turn, converted this torque to a constant contact force between the ultrasonic horn tip and the bottom of the specimen. The load applied to the specimen was measured using the internal system of the Instron testing machine. The elongation was monitored with an LVDT attached to the specimen as shown in Fig. 2 (N).

A quartz transducer (O), possessing a resonance frequency of 8 MHz, was coupled to the flat upper end of the specimen with a light oil couplant and clamped in place. The clamping fixture (P) maintained constant contact pressure between the transducer and specimen end face. The quartz transducer was electrically connected through an impedance matching network to a Matec[b] ultrasonic pulse-echo system. Both longitudinal and shear quartz transducers were used. The propagation velocity

[a]Instron Engineering Corp., 250 Washington Street, Canton, Mass. 02021.
[b]Matec, Inc., 60 Montebello Rd., Warwick, R.I.   02886

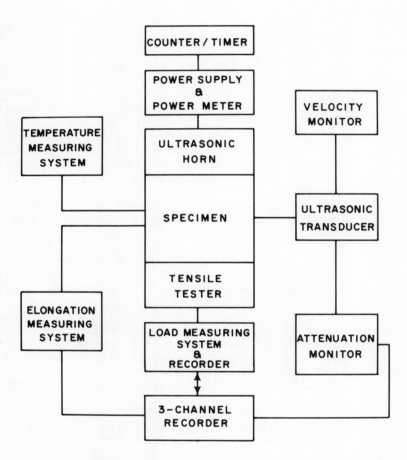

Fig. 1.   Block Diagram of Experimental Instrumentation

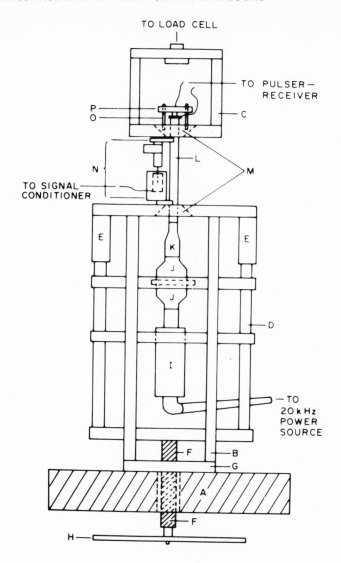

Fig. 2.   Schematic Diagram of Apparatus
          A-moving crosshead; B-bottom grip cage; C-upper
          grip cage; D-mount for ultrasonic horn converter
          assembly; E-guide tubes; F-horn positioning
          screw; G-lower plate of bottom grip gage;
          H-pulley; I-converter; J-1:1 booster; K-cate-
          noidal horn; L-specimen; M-split conical clamps;
          N-LVDT, fixture and positioner; 0-8 MHz trans-
          ducer; P-transducer holder.

of the elastic waves was measured using the ultrasonic
pulse-echo-overlap technique originally described by
Papadakis[4],[5] as modified by Chung, Silversmith, and
Chick[6]. Measurements of elastic wave velocity directly
yield information on the variation of the effective
Young's modulus as a function of deformation and insonation. The same system permitted simultaneous measurements of the ultrasonic attenuation by addition of a
Matec Automatic Attenuation Recorder. The ultrasonic
attenuation measurements served as a monitor of dislocation motion during mechanical deformation and insonation.

In earlier experiments, the temperature of the test
specimen was monitored by cementing a copper-constant
thermocouple to the central region of the specimen gauge
length. Later experiments were performed while monitoring the test specimen temperature with an AGA Thermal
Imaging System[c]. The thermal imaging system incorporated
both a black and white gray scale monitor and a color
monitor. The color monitor made it possible to quantify
the relative thermal effects produced by the ultrasonic
horn. The minimum resolvable temperature difference was
0.2°C using the color-coded isotherms of the color
monitor. The scan rate of the system was 16 frames per
second. A 35mm camera was used to obtain permanent
records of the thermal images throughout the test.

RESULTS AND DISCUSSION

The data presented in Fig. 3 was obtained from an
aluminum single crystal 1.25 cm diameter and 10.60 cm in
length. Displayed are the load, elongation, percentage
change of wave speed or ultrasonic velocity, and the
change of ultrasonic attenuation, all shown as a function
of time. The onset and duration of insonation are indicated by the vertical dashed lines labeled with Roman
numerals.

A similar set of data for another aluminum single
crystal, 1.25 cm in diameter and 10.68 cm in length, is
shown in Fig. 4. The results of both tests were quite
similar for equivalent or nearly equivalent insonation
periods. There appeared to be some difference in the
effects of the length of the insonation period on the
load, elongation and percent velocity change, but only
in magnitude. Sharp, rapid load drops were observed at
the onset of each insonation period in both crystals.

---

[c]AGA Corp., 550 County Ave., Secaucus, N.J. 07094

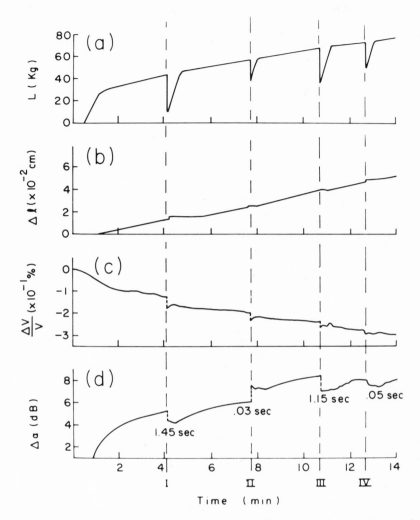

Fig. 3. Tensile load (L), change in specimen length
(Δℓ), percent ultrasonic velocity change (ΔV/V)
and change in ultrasonic attenuation (Δα) as a
function of time for 20kHz high-power insonation
superimposed for various time intervals: I-1.45
sec.; II-0.03 sec.; III-1.15 sec.; IV-0.5 sec.
during tensile elongation of a 1.25 cm diameter
10.60 cm long aluminum single crystal specimen.

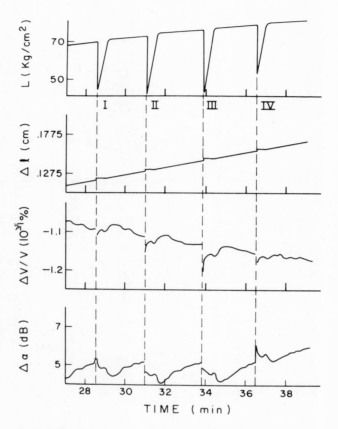

Fig. 4.   Tensile load (L), change in specimen length (Δℓ), percent ultrasonic velocity change and change in ultrasonic attenuation (Δα) as a function of time for 20 kHz high-power insonation superimposed for time intervals of: I-0.03 sec.; II-0.33 sec.; III-0.33 sec.; IV-0.03 sec. during tensile elongation of a 1.25 cm diameter 10.68 cm long aluminum single crystal specimen.

These load drops were essentially similar to those
reported by previous investigators[2]. Accompanying each
load drop was a rapid increase in the specimen length.
At a chart recorder speed of 2 cm/min, these increases
in length appeared to occur immediately. At a chart
speed of 120 cm/min there was no delay observed between
the onset of insonation and the start of elongation.
Although the Instron tensile test machine crosshead speed
during these tests was 0.005 cm/min, a strain rate of the
order of 2.5 cm/min was measured during an 0.05 sec
insonation period. The elongation was observed to
increase linearly with insonation time and continued in
a much slower nonlinear fashion for approximately 0.5
sec following cessation of insonation.

     In both specimens, insonation caused a rapid de-
crease in ultrasonic velocity indicative of alterations
in the elastic moduli of the metal specimens. However,
since the magnitude of the observed velocity decrease
was only of the order of 0.1% or less, the associated
change in elastic moduli would also be small and could
therefore not be responsible for the large load drop
observed during insonation.

     The ultrasonic attenuation exhibited a much greater
change than the velocity as a result of insonation. For
single crystal aluminum specimens subjected to simultaneous insonation and tensile elongation, the attenuation
was observed to increase sharply for insonation times of
the order of 0.03 seconds and then to decrease gradually
almost back to the initial value following cessation of
insonation. Insonation periods longer than 0.03 seconds
caused the attenuation to decrease sharply and then to
increase gradually almost back to the initial value
following cessation of insonation.

     The effects of insonation on the load and elonga-
tion for polycrystalline aluminum specimens were similar
to those observed for the single crystal aluminum
specimens. However, the effects of insonation on the
velocity and attenuation were somewhat different. The
attenuation increased for the longer insonation periods
and decreased for the short insonation period. The
velocity changes were found to be opposite of the attenuation changes, decreasing for long insonation times and
increasing for short times.

     For all of these tests the ultrasonic attenuation
and velocity went through various undulations, which
appeared to be caused by an interplay between the

magnitude of the load drop and reloading of the specimen
by movement of the tensile test machine crosshead. In
an attempt to separate the interplay, two procedures
were used. A simulated load drop was produced by a rapid
unloading of the specimen via manual control of the
Instron and subsequent reloading as would normally occur.
The other procedure was to monitor the parameters during
insonation of specimens not subject to tensile loading.
The results of the simulated load drops are illustrated
in Fig. 5. The simulated load drops appear quite similar
to those caused by insonation; however, no elongation
occurred, the velocity increased initially then decreased
and the attenuation change was somewhat different.
Typical results obtained from a specimen insonated while
no load was applied are shown in Fig. 6 for various
insonation periods. In this case the ultrasonic attenuation was observed to increase sharply upon initiation of
insonation and then to monotonically decrease back to the
initial value upon cessation of insonation. No undulations in attenuation were observed in these tests as
they were when tensile elongation and insonation were
superimposed. A decrease in velocity was observed for
all insonation periods followed by some recovery. The
manner in which both attenuation and velocity change in
specimens subject to insonation only appear to be independent of the length of insonation time. Some heating
was observed for insonation periods longer than about
1.25 seconds and a subsequent thermal expansion was
observed in the specimens as indicated in Fig. 6.

The specimen temperature was initially monitored
using copper-constantan thermocouples epoxied to the
surface. The thermocouples indicated that the temperature of the specimen increased very rapidly during
insonation and then decreased exponentially to its
original value upon cessation of insonation. The magnitude and manner of the temperature increase was independent of the position of the thermocouple along the
length of the specimen. However, touching the specimen
with a finger after insonation revealed that the specimen
was not as hot as indicated by the thermocouple and that
a temperature gradient did appear to exist along the
specimen length. This contradiction in observations was
resolved by monitoring the temperature distribution on
the surface of a number of test specimens during high-
power insonation using an AGA infrared thermovision
imaging system at Virginia Polytechnic Institute and
State University. Use of the AGA thermal imaging system
during insonation revealed an extremely rapid temperature
increase at the point of attachement of the thermocouple

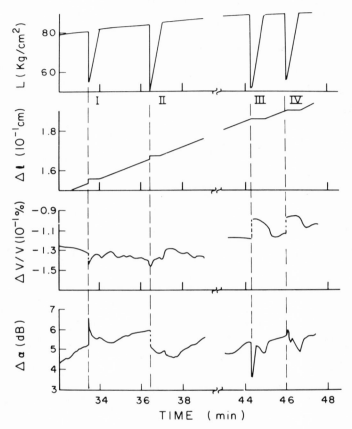

Fig. 5. Tensile load (L), change in specimen length (Δℓ),
percent ultrasonic velocity change and change in
ultrasonic attenuation (Δα) as a function of time
for 20 kHz high-power insonation superimposed for
time intervals of: I–0.03 sec. and II–0.33 sec.
compared to the effects on the parameters when
the aluminum single crystal (1.25 cm diameter;
10.68 cm long) was unloaded by hand; III and IV.

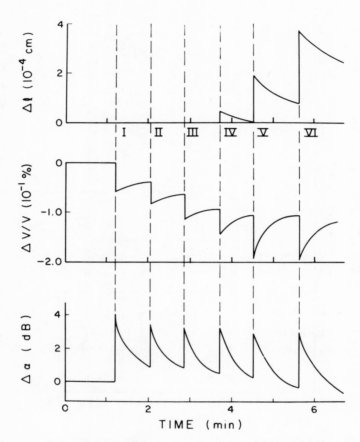

Fig. 6.   Change in specimen length (Δℓ), percent ultrasonic velocity change (ΔV/V) and change in
ultrasonic attenuation (Δα) as a function of
time for 20 kHz high-power insonation superimposed for time intervals of: I-0.03 sec.;
II-0.63 sec.; III-1.23 sec.; IV-1.83 sec.;
V-2.43 sec.; VI-3.63 sec. on an aluminum single
crystal (1.25 cm diameter 10.73 cm long) not
subjected to a static load.

to the test specimen, shown schematically in Fig. 7.   If
the thermocouple were removed, no local heating was
observed except at the horn-specimen contact area.
Heating at the thermocouple was found to be independent
of location of the thermocouple on the specimen surface
and also independent of specimen length.   The heat
generated at the thermocouple contact area flowed into
the specimen, but the temperature increase was not as
rapid nor as great as the thermocouple indicated.   This
extremely rapid temperature increase at the point of
attachment of the thermocouple casts strong doubt on all
previous temperature measurements made using thermocouples.   It should be noted that the many other investigators have relied on such thermocouple temperature
measurements.

        For very short insonation periods (less than 0.03
sec) there was no observed change in specimen temperature.   However, for longer insonation periods various
amounts of specimen heating were observed depending upon
the specimen geometry and test configuration.   As shown
schematically in Fig. 8 specimens of non-resonant
length when subjected to both tensile loading and insonation showed only moderate heating which originated at
the horn-specimen contact area.   On the other hand, one-
half wavelength long resonant specimens of brass, copper,
and steel were observed to heat up to 1000°C in approximately 20 sec at their nodal point, illustrated in
Fig. 9.   Similar specimens of aluminum were observed to
heat up much slower.   These resonant specimens were
not subjected to tensile loading and were threaded into
the horn tip.   Many of these specimens experienced
fatigue fracture at the nodal point after a few minutes
insonation.

        Tests run on large-grained polycrystalline zinc
specimens revealed rapid localized heating at certain
grain boundaries.   Insonation induced severe bending
of one test specimen similar to that observed by Langenecker[7].   Tests run on fine-grained polycrystalline
specimens possessing artificially induced defects such
as saw-cuts and drilled holes revealed insonation
induced hot zones at the defect locations.   Similar
results were obtained on specimens possessing fatigue
cracks.

SUMMARY

        At the onset of insonation a sharp, immediate load
drop was observed, similar to that reported by other

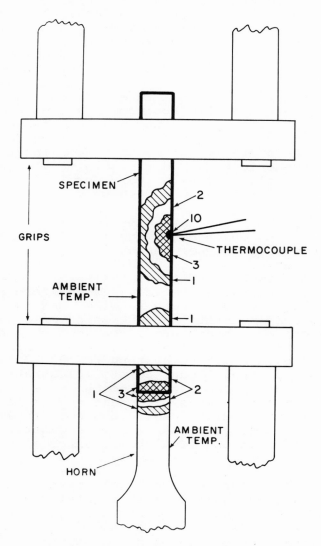

Fig. 7.   Schematic diagram of the temperature gradient
          observed in an aluminum single crystal test
          specimen with a thermocouple cemented near the
          center of the gauge length.   Reproduced from a
          photograph of the AGA Color Monitor approximately
          four seconds after the onset of insonation.   The
          numbers indicate the temperature above ambient
          in degrees centigrade.

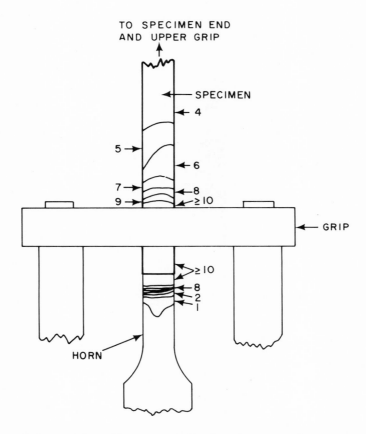

Fig. 8.   Schematic diagram of the temperature gradient
observed in an aluminum single crystal test
specimen without a thermocouple. Reproduced
from a photograph of the AGA Color Monitor
approximately two seconds after the onset of
insonation. The numbers indicate the temperature above ambient in degrees centigrade.

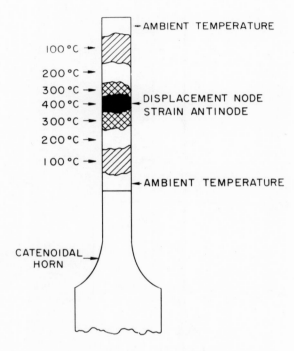

Fig. 9.   Schematic diagram of the temperature gradient
          observed in a one-half wave length resonant
          specimen of steel. Reproduced from a photo-
          graph of the AGA Color Monitor approximately
          10 seconds after the onset of insonation.

workers. At the onset of insonation a sharp, immediate specimen length increase was measured. Insonation caused a rapid change in ultrasonic attenuation indicative of dislocation motion. Insonation caused a rapid change in ultrasonic velocity indicative of alteration in the elastic moduli of the metal specimens. The changes in ultrasonic attenuation and velocity were different for single-crystals and polycrystalline specimens. The changes in ultrasonic attenuation and velocity were different for insonation times of the order of 0.03 sec than for longer insonation periods. Infrared scans of the test specimens indicated that the observed changes were not caused by specimen temperature increases for short insonation times. For long insonation periods some thermal influence must occur due to the increase in temperature of the specimens. Infrared scans revealed an extremely rapid temperature increase at the point of attachment of thermocouples to the test specimen's surface. This result casts strong doubt on all previous temperature measurements since the vast majority of prior investigators used thermocouples. The heat flow in single-crystals and fine-grained polycrystalline aggregates was similar, with the heat flowing uniformly into the specimens from the horn contact end for nonresonant specimen lengths, and with a hot zone developing preferentially at the center of specimens possessing resonant lengths. Tests run on large-grained polycrystalline specimens possessing artifically induced defects such as saw-cuts and drilled holes revealed insonation hot zones at the defect locations. Similar results were obtained on specimens possessing fatigue cracks.

ACKNOWLEDGMENTS

This material is based upon work supported by the National Science Foundation under Grants No. DMR-76-02537 and DMR-78-05994. A special note of thanks in this regard is due Robert J. Reynik. The authors wish to express their gratitude to Alexander G. Rozner, Edmund G. Henneke, II, John C. Duke, Jr., Kenneth L. Reifsnider, and Jeffrey Glass for their technical assistance. The authors would also like to thank Alicia Falcone for typing the manuscript.

REFERENCES

1. F. Blaha and B. Langenecker, Dehnung von Zink-Kristallen unter Ultraschalleinwirkung, Naturwissenschaften, 42:556 (1955).

2. R. E. Green, Jr., Non-linear effect of high-power ultrasonics in crystalline solids, Ultrasonics, 13:117 (1975).

3. R. B. Mignogna and R. E. Green, Jr., Multi-parameter system for investigation of the effects of high-power ultrasound or metals, Rev. Sci. Instrum., 50:1274 (1979).

4. E. P. Papadakis, Ultrasonic attenuation and velocity in three transformation products in steel, J. Appl. Phys., 35:1474 (1964).

5. E. P. Papadakis, Ultrasonic phase velocity by the pulse-echo-overlap method incorporating diffraction phase corrections, J. Acoust. Soc. Am., 42:1045 (1967).

6. D. H. Chung, D. J. Silversmith, and B. B. Chick, A modified ultrasonic pulse-echo-overlap method for determining sound velocities and attenuation of solids, Rev. Sci. Instrum., 40:718 (1969).

7. B. Langenecker, W. H. Frandsen, and S. Colberg, Kinking in zinc crystals by ultrasonic waves, J. Inst. Met. 91:316 (1963).

THE MECHANICS OF VIBROTHERMOGRAPHY

K. L. Reifsnider, E. G. Henneke, W. W. Stinchcomb

Engineering Science and Mechanics Department
Virginia Polytechnic Institute and State University
Blacksburg, Virginia 24061

ABSTRACT

Vibrothermography is a term coined by the authors to describe
a concept and related techniques whereby the internal integrity
and uniformity of materials and components is interrogated by observing the heat pattern produced by the energy dissipation which
occurs when a specific vibratory excitation is applied to the test
piece. In particular, "vibrothermography" consists of the study of
thermographic (heat) patterns which are recorded or observed in
real time during such an excitation. It has been observed that the
details of the mechanisms which produce such dissipative heat are
directly related to the mechanisms of material deformation and degradation in several important ways, a fact that provides the basis
for the use of this scheme as a nondestructive test and evaluation
method and general philosophy. The present paper is an overview of
this field which begins with a tutorial explanation of the basic
foundations of the method, continues with a development of the
manner in which the information provided by the method is related
to the mechanics of the response of the test specimens — to the
extent that it is understood —and closes with a survey of results
and applications of the method.

INTRODUCTION TO THERMOGRAPHY

Thermography refers to the mapping of isotherms, i.e., contours
of equal temperature, over a surface. Both contact (e.g., liquid
crystals) or noncontact (infrared radiometer) methods may be used to
obtain the temperature profiles. For the present paper, emphasis
will be placed upon using real time video-thermography as the (noncontact) monitoring method. Real time, video-thermography utilizes

249

an infrared detector in a sophisticated electronics system which conditions the detected infrared signals and displays these upon a tele-
vision monitor. The method is real-time to the point that the displayed pattern can change as rapidly as the framing rate used by the
infrared camera-television monitor system.

The surface isotherms in the material under investigation may
be established either by external heating (sometimes referred to in
the literature as passive heating) or by internally generated heat
(referred to as active heating). In the passive heating case, an
external heat source is applied to the specimen and one looks for
isothermal patterns that display variations because of local differences in thermal conductivity. The variances in thermal conductivity may arise because of local inhomogeneities or flaws in the
material. For active heating, one applies some other energy source,
such as electrical or mechanical energy, and looks for isothermal
patterns established by the transformation of this energy into heat.
Again, local inhomogeneities or flaws will generally transform the
energy in a different manner than the material-at-large, and variations in the isothermal contours will occur. The thermography technique, then, is mainly concerned with differences in temperature,
i.e., the thermal gradients, that exist in a material and not so much
with the absolute value of the temperature.

Noncontact thermography using infrared radiometers is based
upon the fact that all bodies at temperatures above absolute zero
spontaneously radiate electromagnetic energy at frequencies below
that corresponding to the red spectrum of visible light. The spectral
range of the portion of the electromagnetic spectrum that is usually
called infrared ranges from about 10^{15} Hz (just below red) to about
5×10^{11} Hz (just above microwaves). Planck found that the energy
emitted or absorbed per unit surface area per unit increment of wave
length for a blackbody was given by his well-known spectral distribution law:

$$W = 2\pi \ (10^{-9}) \ \frac{hc^2}{\lambda^5} \ [e^{\frac{hc}{\lambda kT}} - 1]^{-1} \tag{1}$$

where W is the spectral emittance (Watt/m^2/mμ), h is Planck's constant
(6.6252 x 10^{-34} J Sec), c is the speed of light (2.99793 x 10^8 m/Sec),
λ is the wavelength, e is the Naperian or natural base of logarithms,
k is Boltzman's constant (1.3804 x 10^{-23} J - °K^{-1}), and T is the
absolute temperature (°K). Planck's law applies only to blackbodies,
i.e., bodies which absorb all of the radiation incident upon them and
emit the maximum possible amount of radiation at any given temperature
Such bodies do not actually exist in nature but may be approximated
in the laboratory by constructing a cavity inside of a body and
measuring the radiation emitted from a small hole in the surface [1].

All real bodies emit radiation which is some fraction of that
given by equation (1). To account for this, a quantity known as the
emissivity is defined and multiplied into this equation. The emissivity is always less than one (being exactly equal to one only for
a blackbody). Furthermore, one may speak of a "total emissivity" ——
the fraction of total energy emitted at a given temperature in comparison to a blackbody —— or a "spectral emissivity" —— the fraction
of energy at a particular wavelength that is emitted. The total
emissivity of a body might be quite low while the spectral emissivity
of certain wavelengths may be nearly unity. Both types of emissivity
may vary with temperature, physical phase, surface finish, molecular
surface layers, etc. The emissivity can not be calculated from any
basic physical model but must be determined experimentally for each
object. It is strictly a surface characteristic for opaque materials.
The emissivity ranges from zero for "mirror-like" surfaces to nearly
one for lamp black, zapon black, and such surfaces [1]).

For an infrared system viewing any source, the received power at
the system aperture is given by [1]

$$H = \frac{W\omega}{\Omega} \qquad\qquad\qquad\qquad\qquad (2)$$

where W is the radiant emittance given by Planck's Law, equation (1),
ω is the angular field of view of the viewing system, defined by the
optical system and the detector, and Ω is the total solid angle about
the source. Hence, it can be seen that the primary parameter which
governs the response of the infrared viewing system is the relationship between the angle subtended by the source compared to the angular
field of view of the system. If the source is small compared with the
field of view of the detector, the received radiation will vary with
distance between the source and detector but not with angle about the
source. Interestingly, if the source is large compared with the field
of view, the received radiation varies with neither distance to the
source nor angle about it. This latter fact is a result of Lambert's
cosine law which says that the radiant energy emanating in a given
direction from any point on a surface is a function of the cosine of
the angle between the normal to the surface at that point and the
given direction. More specifically, the maximum radiation occurs
along the normal direction to the surface and none tangentially to it.

The significance of Lambert's law for infrared detection is that
a detector looking at an emitting surface will always receive the same
amount of energy no matter what the angle between the detector's
line of sight and the radiating surface. This statement is true for
both planar and general, curved surfaces, although it is easier to
see for planar surfaces. As the angle between the line of sight and
the normal to the surface increases, so does the area of the surface
viewed by the detector. The increase in viewed area exactly matches
the reduction in radiation energy as given by Lambert's law so that
the total energy impinging on the detector is constant. This, of

course, is valid only as long as the emitting surface completely
fills the detector's field of view. A simple rule in utilizing a
detector system then is to make the source the only object in the
field of view by appropriately controlling the field stop of the
system.

A secondary consideration that one must be aware of when using
thermography as an NDE method is the fidelity of observed surface
thermal patterns to interior inhomogeneities or flaws. The relationship between the surface isotherms and the interior thermal patterns
is, of course, governed by the thermal conductivity of the material
and the distance between the surface and an interior region of interest. For very thin materials such as composite laminates, it has
been shown that the fidelity of surface thermal patterns to interior
flaws is quite good [2]. For bulk materials, thermography may be
useful for initial indications of the presence of flaws but will not
be nearly as good as other NDE methods for resolution of the flaw
size and shape. For example, what might seem to be a rather sharp
discontinuity may prove to have a weak thermal signature because of
heat conduction through the material [3]. The thermal signature can
be improved by making the cooling rate at the surface as large as
possible. This may be done by making the surface as nonreflective as
possible and by augmenting radiant cooling by forced circulation of
air over the test surface.

There are two basic types of infrared detectors: (1) photon-
effect devices and (2) thermal devices. The photon-effect devices
are sensitive to the wavelength of the received radiation while
thermal detectors respond only to the heating caused by the incident
radiation and are largely independent of the wavelength. Real-time
video-thermography requires that the entire field of view be scanned
very rapidly so that the temperature of each field point is measured
and displayed upon the cathode ray tube. Hence such systems require
the high sensitivity and fast response time of photon-effect devices.

Photon-effect devices are simply composed of materials which
produce voltage, current, or resistance changes when irradiated by
photons. They are generally classified as photoemissive detectors,
photoconductive detectors, or photovoltaic detectors. Photoemissive
detectors, discovered by Hertz, are materials which emit electrons
when irradiated by photons having wavelengths less than a critical
value. These devices are primarily responsive through the visible
wavelengths down to approximately one micron and hence are not
extensively used in infrared systems. Photoconductive materials are
semiconductors whose conductance change when irradiated by photons.
Their response time is very fast with times shorter than one micro-
second having been reported. For maximum sensitivity, it is often
necessary to cool the semiconductor material to reduce thermal noise.
Photovoltaic cells are composed of p-n junction semiconductor materials. The cell produces a voltage when irradiated by photons.

Several commercially available thermographic systems currently use photoconductive detectors which must be cooled to liquid nitrogen temperature for proper operation. One can also purchase detectors which have maximum sensitivity to different wavelengths and thus choose a system which has optimum capability in the temperature range of interest.

As mentioned earlier in this section, in order to establish surface thermal gradients which can be detected by an infrared system, it is necessary to use either a passive or active heating method. The present authors have found that an active heating method is more advantageous when trying to determine, through an NDE method, an indication of damage that may influence mechanical response of the material. Two methods of active heating have been employed in our laboratory — large amplitude, low frequency mechanical (fatigue) loads and small amplitude, high frequency mechanical vibrations. In the former case, thermography can be used as a real-time monitor of damage initiation and growth while a specimen is being fatigued. In the second case, the amplitudes of the mechanical vibrations are so small that mechanical damage is not induced in the specimen by the presence of the vibrations. However, the transformation of mechanical to heat energy is preferential around the regions of stress concentrations induced by damage already present in the material. The latter technique has been named vibrothermography [4] because of the need for the combined use of mechanical vibrations and videothermography to make the technique viable. A discussion of the phenomena responsible for mechanical-to-heat-energy conversion and examples of applications of these techniques are given in the subsequent sections.

THE MECHANICS OF HEAT GENERATION

If the response of a material to loading (or other applied environments) is entirely linear, i.e., the response is single valued for a fixed input regardless of prior loading history, no energy loss would result and no corresponding heat would be produced. Hence, the dissipative mechanisms responsible for heat generation during loading which form the basis of the vibrothermography technique are due to some departure from this strict linearity. In that context, it is interesting to note that most of the literature which deals with these mechanisms is based on analytical representations which introduce a damping or dissipative term into a linear theory by such means as assuming that the elastic stiffness has a real and imaginary part, or that there is a phase lag between stress and strain, or some other scheme related to damped oscillator theory. As a consequence, it is commonly assumed that superposition of influences applies, that the specific dissipation of a material does not depend on the amplitude of the excitation, and that other assumptions familiar to linear treatments hold.

A number of physical dissipative mechanisms appear to operate in

such a way that these linear representations are appropriate [5].
At low stress levels anelastic effects including the various relaxation phenomena (atomic diffusion, order-disorder processes, thermal
diffusion, etc.) are common. Over a larger range of stress, viscoelastic effects are observed. While the mechanics of these two types
of behavior are closely related, anelasticity has developed as a
rather special part of the subject having to do with physical and
chemical phenomena in crystalline solids. For the most part, the
object of that development was, and is, to study the material itself
as a bulk substance, and to establish the character of certain internal features by their relaxation spectra. Viscoelasticity is an
older and more general field having closer ties to engineering situations wherein the inherent response of a material to an applied
state of stress or strain is rate dependent. Viscoelastic creep
under quasi-static or static loading is a common consequence of this
type of behavior although cyclic response is also an active part of
this sub-field.

From our present point of view, some of the dissipative mechanisms associated with anelasticity are not of practical importance.
As a group, this is especially true of the atomic diffusion events
that are activated by stress, including such things as grain boundary
motion, single or paired solution atom motion, and twin boundary
activity. While these events may be prominent in the small strain
range, the heat produced by them is not significant because of the
low ratios of dissipated-to-input energy, the low levels of input
energy (for small strains) and the low frequencies at which these
events are commonly excited [6].

A major part of the classical field of engineering damping, as
associated with noise control and vibration of structures, is also
based on linear concepts, including a large dependence on viscoelastic damping as a common mechanism. (c.f. references 7-9.)
"Viscoelastic" effects and "anelastic" effects include, as noted
earlier, several other mechanisms such as grain boundary or phase
boundary effects, eddy current effects, and various molecular motions
in long-chain-molecule materials. Only the last of these is of
practical importance in the present context. These types of dissipative mechanisms can produce relative damping values of 0.1 to
1.5 [5]. Furthermore, since long-chain (polymeric) materials are
generally poor conductors, temperature changes at local sources are
generally higher and easier to observe in those materials. In
composite materials, or in other situations where there is an oscillating nonuniformity of stress, thermal currents due to local variations in temperature produced by anelastic response can also produce
dissipation.

Anelastic effects are nearly always present. One major reason
for this is that the magnitude of temperature changes due to adiabatic straining at practical excitation frequencies is often

significant, especially for the more conductive materials. The
temperature change for adiabatic elastic deformation is given by [10],

$$\Delta T = -\frac{3T\alpha Ke}{\rho c} \tag{3}$$

where α is the coefficient of thermal expansion, K is the bulk
modulus, e is the volume dilatation, ρ is the mass density and c is
the specific heat of the material being strained. For aluminum
loaded over a range between the tensile and compressive yield
strengths in a uniaxial test at 1 Hz the entire specimen may oscillate over a temperature range of 0.5°C or more, depending upon the
type of aluminum, the conductive, convective and radiation losses and
other experimental factors. Such large temperature oscillations
which cycle in phase with the excitation are easily observed in many
materials and have been observed for several metals and plastics in
our laboratory. However, for homogeneous materials the dissipative
loss associated with thermal currents induced by these temperature
changes is generally rather small. In general, the losses are proportional to the difference between the adiabatic and isothermal
elastic compliances of a given material, i.e.,

$$S_{ij}\Big|_{\text{adiabatic}} - S_{ij}\Big|_{\text{isothermal}} = -\frac{\alpha_i\alpha_j}{c}T \tag{4}$$

where S_{ij} are the elastic compliance tensor components, α_i are the
vector components of thermal expansion, T is the temperature and c is
the specific heat of the material. The physical reason for the
strength of the dissipative loss being controlled by expression (4)
is simply the fact that the difference between the adiabatic and
isothermal behavior controls the maximum area in the ideal hysteresis
loop associated with what is commonly called the "elastic after-
effect," shown in Fig. 1. As an example of this situation, for
"transverse thermal currents in a reed" (beam) of rectangular cross
section Nowick gives the internal friction as [11],

$$\tan \delta = \frac{E\alpha^2 T}{c} \left[\frac{\omega\tau}{1 + \omega^2\tau^2} \right] \tag{5}$$

where E is the elastic modulus, ω is the frequency of excitation, α,
T and c are defined in Eqn. (3) and the relaxation time is

$$\tau = \frac{a^2}{\pi^2 D} \tag{6}$$

where a is the reed thickness and D is the thermal diffusivity.
However, it should be emphasized that these expressions are for
homogeneous materials. Virtually no basic characterization of
thermal current loses in inhomogeneous materials such as composites
has been carried out. Values of relative damping given by Eqn. (5)

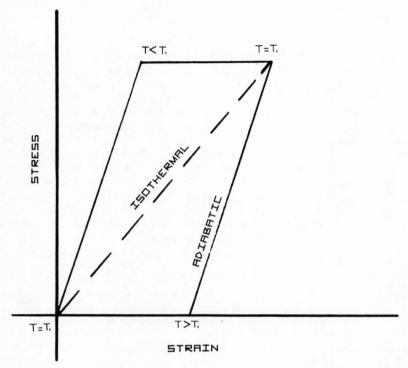

Fig. 1. Idealized hysteresis loop due to elastic after-effect.

usually fall in the range of 0.001 to 0.01, small by engineering
standards.

Figure 2 shows an interpretation of this equation for some materials of interest for a range of region sizes of practical importance. Two types of practical information are illustrated by Fig.
2. If energy dissipation is due to thermal currents within the
region of interest, then the frequencies of oscillation for a range
of materials which will maximize the dissipative loss is given
directly by the figure. It is important to notice the strong dependence on material properties in Fig. 2. (The ordinate is a loga-
rithmic scale.) The material property that contributes the most to
this dependence is the thermal conductivity. The density and specific
heat rarely change by an order of magnitude but the thermal conductivity may change by four (or more) orders of magnitude. This

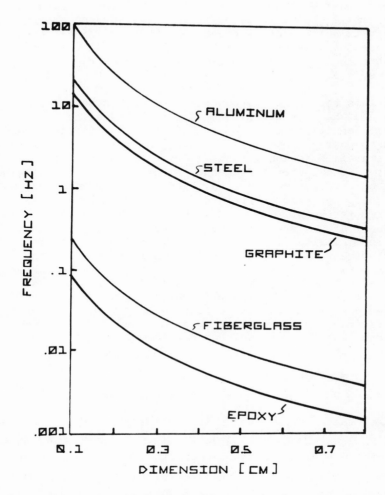

Fig. 2. Region size dimension effect on relaxation frequency.

will be a recurring situation as we continue our characterization:
the thermal conductivity has a dominant influence over the nature of
the heat patterns in comparison to all other physical variables such
as surface film coefficients, emissivities or geometry effects, given

a heat source of fixed intensity. It should also be noted that for regions of sufficient size to be readily observed as having a sharp difference in temperature, Fig. 2 indicates that excitations in the ranges below 1 Hz are required for materials which are poor conductors. If we consider frequencies which are significantly higher than those shown we can assume that the regions of interest have constant temperature distributions since conduction is not a major influence, i.e., this is essentially an adiabatic condition.

In this regard, a few observations can be made concerning the evolution of heat by stress waves, either traveling or standing. The adiabatic condition is not quite satisfied for longitudinal waves since heat energy flows from the hot, compressed regions to the cooler, rarefied regions. This interchange of heat energy will, by necessity, remove energy from the mechanical wave, causing general heating of the body and eventual extinction of the stress wave. One might expect there to be preferential heating due to this effect in regions where the stresses are intensified due to material discontinuities. Upon closer examination, however, it is found that this effect is minimal at frequencies of practical interest. This heat loss will occur most readily at frequencies for which the wave length becomes small enough that the compressed and rarefied regions are so close that heat exchange can most easily take place. In fact, for very high frequencies, one approaches an isothermal state where heat exchange takes place nearly instantaneously. In this region, the heat loss from the stress wave is a maximum. The frequency at which this transition from the adiabatic to the isothermal condition occurs can be estimated from [12]

$$f \gg \frac{1}{2\pi} \rho \frac{c_v v^2}{K} \tag{7}$$

where K is the bulk modulus, ρ is the mass density, c_v is the specific heat at constant volume and v is the longitudinal phase speed. For quartz, for example, the transition frequency is of the order of 10^{11} Hz. Hence in all practical cases, this transition will never be observed. The wave attenuation coefficient for this phenomenon in normal frequency ranges has been estimated to be [12].

$$A = \frac{2\pi^2 f^2}{\rho v^3} \left[\frac{K}{c_v} \frac{C_{11}^{\sigma} - C_{11}^{\theta}}{C_{11}^{\sigma}} \right] \tag{8}$$

where f is the frequency, and C_{11}^{σ} and C_{11}^{θ} are the isentropic and isothermal elastic moduli, respectively. For quartz, this value is of the order of 12×10^{-4} nepers/cm for f = 100 MHz. Hence, only at the very high frequencies normally used in ultrasonic testing does this heat generation effect become non-negligible.

For practical situations, most of the interest is in an intermediate frequency range, especialy in view of the results of earlier
work which shows that heat images of defects in several materials can
be observed at frequencies of excitation between about 30 Hz and the
mega-hertz range [2,13,14]. For these frequencies, especially in the
kilohertz range, the wavelength of traveling stress waves in most
engineering materials is greater than the order of a centimeter. We
can then assume that the stress or strain due to the external excitation is uniform over the regions of interest (1-10 mm). Situations
for which the wavelength is of the order of the inhomogeneities or
discontinuities produce what is often called a loss to propagating
stress waves. One can refer, for example, to scattering losses
incurred in polycrystalline materials when the wavelength is comparable to the grain size [12]. This, however, is only an apparent loss
due to the scattering of wave energy in all directions, and hence
away from the primary propagation direction. This loss does not
directly result in a preferential transformation of mechanical to
thermal energy at the scattering site, at least as far as presently
accepted physical models predict.

While the energy dissipated by traveling waves is not large
enough to create observable heat patterns under most practical
circumstances, standing wave patterns are quite capable of heat
generation at practical levels. Forced oscillation of resonant
systems can produce energy levels sufficient to cause outright
fracture, not only of structures, but of material coupons. This
subject is addressed in detail by another paper in this volume, so
that a detailed discussion here would be redundant. In general, the
maximum temperature increases can be expected in the maximum strain
positions, the nodal points in a one dimensional vibration, for
example. Since the mechanics of body vibrations is a rather well
developed field, heat patterns due to that type of excitation can be
anticipated (or avoided). However, it is also possible for "local
vibrations" of material phases or defect surfaces to produce heat.
This type of heat generation, and the corresponding mechanics, is
not well understood. It is likely, in fact, that many of these
mechanisms have not been identified.

One particular type of mechanism which is related to local
vibrations is the "clapping" or rubbing of defect surfaces. When a
flaw such as a crack, a debond or a delamination forms in a material,
new surfaces are formed. Those surfaces usually remain close to one
another, and sometimes remain in contact. When the region of the
flaw is dynamically excited, the surfaces may move independently of
one another and may encounter one another during dissimilar relative
motion by either rubbing (due to relative shear) or clapping (due to
relative motion normal to the surfaces). There is growing evidence
that this type of heat production is very common and very efficient
in materials with complex microstructure and complex damage development modes.

One of the most efficient heat production mechanisms is plastic deformation whereby slip occurs creating permanent shape changes globally or locally. Most of the energy which is required to cause such plastic deformations is dissipated as heat. Such heat development is easily observed at the tip of a propagating crack even in excellent conductors such as aluminum [15]. This type of heat generation mechanism is unique in the sense that it is a very efficient heat producer and it requires "permanent" deformation. It is possible for this mechanism to produce heat in a fixed location by reversed or alternating slip due to cyclic (reversed) applied loads, but such situations do not appear to be common. Large strain nonlinear deformations are, however, usually highly dissipative by their very nature so that if such mechanisms are present they will usually produce very strong heat images.

A final mechanism should be added. Our discussion above has been primarily concerned with steady-state dissipation, i.e., with mechanisms which produce heat continuously under continuous excitation without changing the local condition of the material. (Of course, plastic slip is an exception to this generality.) Heat is also produced when fracture surfaces are formed. This transient type of heat release can be large and easily observable and can be used as the basis of a method to detect damage development. Since our emphasis here is on the use of vibrothermography as a nondestructive method of observation we will not dwell on this aspect of heat generation in the present paper.

In summary, then, a very large number of dissipative mechanisms have been identified in materials and material systems. Only a few of them appear to be capable of producing heat patterns which can be detected under excitations which are readily available over frequency ranges of interest to most investigators — commonly between about 1 Hz and a few mega Hz. Of those mechanisms, viscoelastic dissipation, standing wave resonant dissipation, plastic slip, and the rubbing and clapping of adjacent surfaces appear to be the only sources of energy which commonly produce the heat patterns observed by vibrothermography techniques.

OBSERVABILITY OF HEAT PATTERNS

We consider a local source of heat with fixed strength (fixed energy flux per unit time out of that local region of fixed size). Then we ask "what are the variables which control our ability to observe and record heat patterns in the neighborhood of that heat source?" It is especially important to understand how material properties affect the pattern and its observability since we need to be able to anticipate the usefulness of vibrothermography for various materials in various situations. The limitations due to observational equipment are usually fairly clearly stated by the builder and are common to all applications using that instrument, generally speaking.

We concentrate, then, on the relationship of material properties to the heat pattern produced by a given source, especially as those properties determine the geometric dimensions and thermal intensity of the patterns, the two characteristics which generally control the observability of the patterns. We attempt to set a generic problem by idealizing the heat source as a circular region of radius r_1 in an infinite plate with unit thickness. For such a situation, the temperature distribution is given by the classical solution of Fourier's equation,

$$(T(r) - T_1) = \frac{Q}{2\pi K} \ln \frac{r}{r_1} \tag{9}$$

where K is the conductivity and Q is the source strength. A tabulation of temperature difference per unit source strength is listed in Table 1. The thermal conductivity controls the behavior, of course, but the nonlinear dependence on the distance, r, introduces some interesting behavior. First, as expected, the magnitude of the temperature difference for insulating materials is three orders of magnitude greater than for good conductors. Second, the sharpness of the temperature gradients is also different, i.e., for good conductors the distance over which the temperature change is spread is less. Both of these factors affect the acuity of the heat image. The temperature difference must be of the order of 1°C for most of the optical image forming thermographic instrumentation available for engineering purposes, especially the video devices of current interest. For a source of, say, 10 BTU per hour (175.8 watts) which is a value which could be typical for local plastic flow at a defect in a metal, from Table 1, materials with conductivities greater than or equal to steel or graphite would be at or below such instrument sensitivity. The temperature gradient widths must also be greater than the spatial resolution of available equipment. If an imaging camera can resolve 150 lines (75 line pairs in common optical terminology) across its imaging plane, the lenses and optical paths then limit the resolution for a given field of view. For a common video camera, for example, a field of view of about 5 cm might result in a resolution of about 0.33 mm or about 1.5 line pairs per millimeter.

All of the data in Table 1 can be collapsed onto a single curve which simplifies the interpretation of the results. Such a plot is shown in Fig. 3. The abscissa in Fig. 3 represents the percent difference in the temperature at any two points separated by a change of 0.1 in the normalized distance (r/r_1) from the source. The ordinate is the normalized distance itself. Hence, (as shown on the figure) if the instrument used to observe a heat pattern can resolve one percent of the source strength, then the apparent size of the pattern will be about five or six times the dimensions of the actual source (provided that the spatial resolution of the camera is not less than that apparent size). In fact, for many practical situations this

Table 1

r/r_1	Epoxy	Glass-Epoxy	$\Delta T/q$ (°F/BTU) Graphite	Steel	Aluminum
2	1.0	3.8×10^{-1}	7.9×10^{-3}	4.2×10^{-3}	9.3×10^{-4}
3	1.6	6.0×10^{-1}	1.2×10^{-2}	6.7×10^{-3}	1.4×10^{-3}
4	2.0	7.6×10^{-1}	1.6×10^{-2}	8.5×10^{-3}	1.8×10^{-3}
5	2.3	8.8×10^{-1}	1.8×10^{-2}	9.8×10^{-3}	2.2×10^{-3}
7	2.8	1.1×10^{0}	2.2×10^{-2}	1.2×10^{-2}	2.8×10^{-3}
10	3.3	1.3×10^{0}	2.6×10^{-2}	1.4×10^{-2}	3.1×10^{-3}
15	3.9	1.5×10^{0}	3.1×10^{-2}	1.6×10^{-2}	3.6×10^{-3}
20	4.3	1.6×10^{0}	3.4×10^{-2}	1.8×10^{-2}	4.1×10^{-3}
25	4.6	1.8×10^{0}	3.7×10^{-2}	1.9×10^{-2}	4.3×10^{-3}
30	4.9	1.9×10^{0}	3.9×10^{-2}	2.0×10^{-2}	4.6×10^{-3}
35	5.1	1.9×10^{0}	4.1×10^{-2}	2.2×10^{-2}	4.8×10^{-3}
40	5.3	2.0×10^{0}	4.2×10^{-2}	2.3×10^{-2}	4.9×10^{-3}

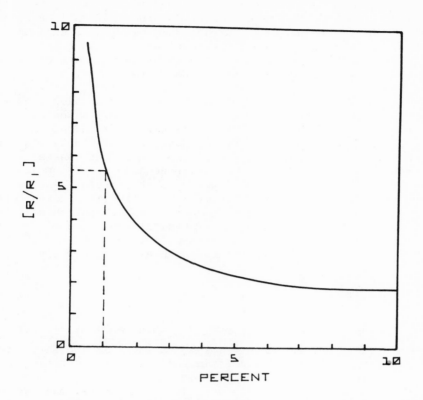

Fig. 3.   Percent difference between successive points on the
          temperature distribution at a source for a change of
          0.1 in r/r_1.

apparent size of five times the source size is a useful rule of
thumb. For example, if we wish to resolve a defect region with a 0.1
in (2.5 mm) characteristic radius and we determine from Table 1 that
our instrument will resolve 1 percent of the temperature range to be
expected, Fig. 3 tells us that the temperature gradient that we can
resolve for this case will appear to be spread over a distance of
about 5.8 times 0.1 or 0.58 in (14.5 mm) which can easily be resolved.
Fig. 3 also indicates that only 10 percent changes will be occurring
at about 0.2 in (5 mm), i.e., most of the apparent heat image will be
quite close to the source.

     We have concerned ourselves with conductivity because of its'
dominant role in the determination of pattern observability. Of

course, the source of heat is just as important since it determines
the maximum level of temperature which is produced. All observational
instruments have a temperature threshold resolution, a minimum
temperature difference which can be resolved. A strong source will
cause a larger pattern and a larger range of temperature differentials
making the source easier to identify and characterize.

It is also possible to have conduction and radiation losses
which significantly influence at least the interpretation of a heat
pattern. The details of such losses are highly dependent on global
details such as specimen geometry, methods of specimen excitation and
absolute temperature levels (for the case of radiant losses). It is
not possible to discuss a representative generic problem which
demonstrates these influences. However, heat transfer is a well
developed subject and should be carefully applied to a given experimental or practical situation. It should also be mentioned in this
context that it is better to measure physical constants if possible
rather than estimate them if wide variability is common. For example,
for convective losses an appropriate balance equation would be:

$$\frac{dT_s}{dt} + \frac{RK}{\rho c} (T_s - T_a) = \frac{q}{\rho c} \qquad (10)$$

where T_s is the surface temperature, T_a is the ambient (or reference)
temperature, R is the surface-to-volume ratio, K is the surface heat
transfer coefficient (or film coefficient), ρ is the material density,
c is the specific heat, q is the source strength and t is time. The
heat transfer coefficient, K, is very difficult to estimate accurately
for a given physical situation. It is easily measured, however, by
a simple experiment, based on the transient (homogeneous) solution of
equation (10). If no heat is supplied (q=0), but an initial temperature difference, $(T_s - T_a)_0$, exists, the surface temperature as a
function of time will be given by:

$$\frac{T_s - T_a}{(T_s - T_a)_0} = e^{-\frac{RKt}{\rho c}} \qquad (11)$$

Hence, by establishing an initial temperature differential (by whatever means) and then measuring the surface temperature as a function
of time, a valid value of K for the surface under study can be determined rather quickly from the cooling curve without any other tedious
calibration. Of course, such "simple" experiments may be complicated
or even invalidated by factors not represented in the balance equation
such as conduction or radiation effects, etc., so that "setting the
problem" correctly is important. However, the important point to be
made is that realistic and more nearly correct values for some
physical constants can and should be measured for the specific cases

of interest when possible.

RESULTS AND APPLICATIONS

     As we mentioned earlier, perhaps the most efficient heat-pro-
ducing mechanism common to engineering applications is plastic deformation (atomic slip) in metals or polymers. An example of such
dissipation is shown in Fig. 4, which is a hysteresis loop produced
by cycling an aluminum bar to equal amplitudes in tension and compression, beyond the yield point in both directions. The energy that
is being dissipated is transformed almost entirely into heat. In the
present case, the aluminum bar, which was 8.9 mm in diameter, reached
a stable average temperature which was about 14°C above room temperature. It should also be mentioned that temperature oscillations of
a few degrees were superposed on that average due to the elastic
after effect. Those oscillations cycled with the frequency of the
applied load (which was about 3 Hz). The material was 6061 Al and
the maximum strain amplitude was about 0.5 percent. The specimen
ruptured after 1420 such cycles.

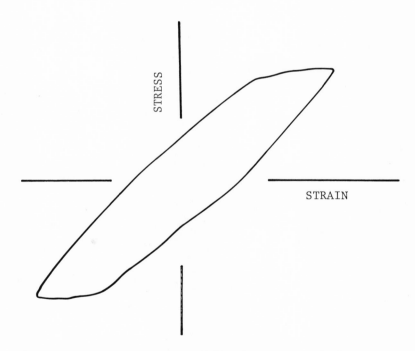

Fig. 4. Hysteresis loop for 6061 aluminum cycled in
        tension-compression fatigue to a strain of
        0.5 percent.

A similar type of behavior can be observed in polymers. Figure 5 shows a closely related but somewhat different result in poly-carbonate. A "dog bone" shaped specimen was extended under quasi-static loading beyond the yield point of that material. Deformation bands formed at the shoulders at each end: one of those bands then began to propagate slowly along the length of the specimen in a very stable way. The nominal specimen thickness and width were 2 mm and 50 mm respectively. After the band had passed a given cross section the thickness changed to 1.6 mm and the width to 45 mm fairly uniformly along the deformed length. During this band propagation the deformation rate of the specimen was held constant and the load was also quite stable. Since the deformation was constrained to the small "gage length" of the deformation band width — which also remained essentially constant — the propagation process was quite close to a steady state situation. This process, then, became of interest, not only because of the physics and mechanics of the phenomenon, but also because it provided us with a well controlled physical situation to use to evaluate our observation techniques and our analytical modelling of those events. A sketch of the propagating

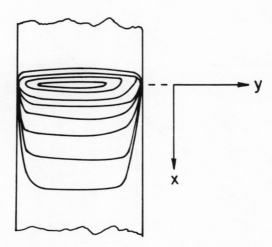

Fig. 5. Propagating heat pattern in a polycarbonate specimen.

heat pattern is shown in Fig. 5. Each of the contour lines represents isotherm, separated by a temperature difference of 2°C, as
recorded by the video-thermographic camera during steady-state deformation band propagation. By counting the contours one can see
that a total temperature increase of 16°C was observed for the band
shown in Fig. 5. The maximum temperature will be influenced, of
course, by the rate of deformation. In the test shown three different
deformation rates were applied. At each rate, the "warming" pattern
was recorded, the steady state temperature pattern was recorded when
the pattern stabilized, and the cooling period was recorded after the
deformation was discontinued. At a crosshead rate of 0.05 mm/min a
stable temperature change of just under 2°C at the highest point in
the pattern was reached. (The initial gage length was about 75 mm
long.) The width of the deformation band zone (the effective gage
length) was observed to be about 5 mm. When the crosshead rate was
increased to 0.25 mm/min a maximum temperature increase of about 9°C
was observed, i.e., an increase of 500 percent in the work rate being
applied to the specimen resulted in a temperature change of almost
exactly that proportion. When the crosshead rate was then changed to
0.5 mm/min the maximum temperature change observed jumped to about
16°C, another nearly proportionate change. Over an order of magnitude
change in deformation rate the maximum temperature change was directly proportional to the work rate. Data were also recorded during
the period between the different rate tests when the specimen was
heating up and cooling down.

    Since the present paper attempts to present an overview, there
is not sufficient space to include a detailed analytical development
of the modelling of heat pattern development. However, the manner in
which the test conditions, specimen geometry and material properties
enter the picture can be suggested by the following abbreviated
exercise. Suppose we consider the temperature pattern sketched (from
a thermogram) in Fig. 5. That pattern was produced by the propagating
deformation band in polycarbonate (described above) for an extension
rate of 0.5 mm/min. The deformation band, and the highest temperature,
are located at the origin of the x-axis on our coordinate system
indicated in the figure. The contours (representing 2°C temperature
steps) are much more widely spread on the lower side of the band than
on the upper side. The lower side corresponds to the region that the
band has just passed over. While most of the deformation is occurring
in the band itself, some plastic flow is evidently still   curring in
the region behind the band, causing heat to be generated over a
length of 35 mm or so. The heat pattern only extends 6 or 7 mm away
from the band on the other side. Since the band was propagating very
slowly under quasi-static loading dynamic effects on the heat pattern
are thought to be negligible.

    For a rough model of the pattern we first observe that the Biot
number for this situation is:

$$Bi = \frac{\bar{h}}{k} \frac{V}{S} \approx 0.2 \tag{12}$$

where $\bar{h}$ is the effective surface transfer coefficient (which is a function of the geometry, ambient properties, and convective flow rate), k is the thermal conductivity of the solid, and V/S is the volume-to-surface ratio [16]. Since the Biot number is small there is little loss in accuracy by considering the problem to be one-dimensional with a line-source at x=0. If the cross-sectional area, A, the specimen perimeter, p, and the conductivity, k, are constant, the Fourier law of conduction and the Newton law of cooling can be combined to yield:

$$\frac{d^2T}{dX^2} - \frac{\bar{h}p}{kA} (T - T_a) = 0 \tag{13}$$

where T_a is the ambient temperature. Equation (13) accounts for loss of heat by conduction away from the source and by convection from the specimen surface across the total heat pattern. Suitable boundary conditions can be set by specifying:

$$T = T_o \text{ at } X = 0 : T = T_a \text{ as } X \to \infty \tag{14}$$

where T_o is the measured maximum temperature at the deformation band. The solution of equation (13) can then be stated as:

$$T - T_a = (T_o - T_a)e^{-mx} \tag{15}$$

where

$$m^2 = \frac{\bar{h}p}{kA} \tag{16}$$

The heat transfer at the band is given by:

$$q_o = \sqrt{\bar{h}pkA} \ (T_o - T_a) \tag{17}$$

Equation (15) predicts the temperature pattern distribution and equation (17) correlates the transfer of heat out of the deformation band with the pattern that develops. The heat transferred out of the band is proportional to the work rate of the forces which extend the specimen (as verified earlier). Some of that work is dissipated in the residual deformtion region below the deformation zone as mentioned above. If we assume, for expedience, that 50 percent of the applied work is dissipated in the deformation band itself, we can use expression (17) to estimate the transfer coefficient, $\bar{h}$. From our measured values of work rate, temperature difference, and geometrical measurements, and using

$$k = 0.00045 \ \frac{cal}{sec \ cm \ °C} \text{ we obtain a value of } \bar{h} \approx 35 \ \frac{W}{m^2 \ °C}$$

which is somewhat highter than expected but a reasonable value which
serves as a check on the validity of our model. Equation (15)
predicts that the exponential temperature drop to the ambient level
T_a will occur over a distance of about 7.5 mm. This is a distance
similar to that observed on the undeformed (top) side of the recorded
pattern. The sensitivity of the thermographic pattern to the specimen (and source) geometry, the material properties and the surface
transfer properties can be assessed from these equations. The heat
transfer aspects of these problems can become very complex requiring
sophisticated analysis methods in some cases. However, the general
nature of these considerations can be extracted from the above
exercise.

Thermography has been used successfully to investigate the
development of damage in fiber reinforced composite materials.
Results from several of the studies are presented. Yeung [17] and
Gibbins [18] investigated the fatigue response of graphite epoxy
laminates containing internal flaws and used thermography to detect
and monitor the growth of damage in real time. The specimens were
$[\pm 45,0]_s$ and $[90,0]_s$ graphite epoxy laminates with 0.25 inch long
slots in the constrained zero degree plies. The slots were oriented
perpendicular to the zero degree fibers [17] and parallel to the zero
degree fibers [18]. The specimens were loaded in tension-tension
fatigue (R=0.1) at 10 Hz in the load control mode.

Figures 6 and 7 show thermographs of a $[\pm 45,0]_s$ specimen and a
$[90,0]_s$ specimen, each with a flaw parallel to the zero degree fibers,
during cyclic loading. For these specimens, the nominal stress in
the constrained zero degree plies is the same and is equal to 85 percent of the tensile strength of an unconstrained zero degree specimen
with a through-the-thickness notch parallel to the fibers. Early in
the cyclic loading history, the temperature of the material at the
flaw site increased. After 12,700 cycles, the thermographic pattern
for the $[\pm 45,0]_s$ specimen is well developed and shows a temperature
difference of approximately 2°C on the surface of the specimen. An
ultrasonic C-scan of the same specimen, made after 10,000 cycles,
shows the size of the flaw which had grown from its original size.
The thermograph after 276,000 cycles is similar to the previous one
in that the difference in surface temperature is 2°C, the hottest spot
is at the site of the flaw, and the size of the hot spot is only
slightly greater than at 10,000 cycles. The thermograph also shows
a different isotherm along the edges of the specimen where the C-scan
indicates delaminations. The C-scan also shows a defect, most likely
a delamination, running diagonally across the specimen; however, the
defect is not indicated on the thermograph. A possible reason for
this will be presented in a later paragraph.

The thermograph of the $[90,0]_s$ specimen shows a typical, well

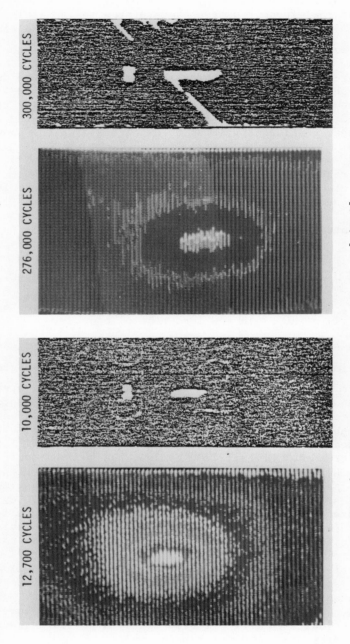

Fig. 6. Thermographs and C-scans of a $[\pm45,0]_s$ graphite epoxy specimen with internal damage.

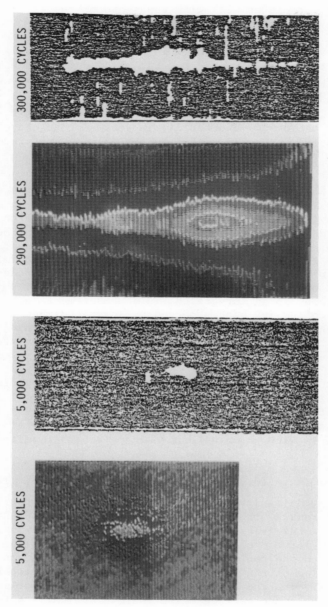

Fig. 7. Thermographs and C-scans of a $[90,0]_s$ graphite epoxy specimen with internal damage.

developed heat pattern with a difference in surface temperature of
0.8°C after 5000 cycles. The corresponding C-scan shows little
damage. At 290,000 cycles, the thermograph shows an elongated
pattern which correlates well with the shape of the flaw indicated in
the 300,000 cycle C-scan.

Further studies of the damage in these specimens by X-radiog-
raphy and sectioning indicated that the major damage modes in the
specimens were delamination and matrix cracks in the zero degree
plies. For conditions of equal fiber direction stress in the zero
degree plies, the $[\pm 45,0]_s$ specimens suffered less damage (smaller
area of delamination and shorter matrix cracks) than the corresponding $[90,0]_s$ specimen, as indicated by the thermographs recorded
during the fatigue tests.

Several analytical investigations have been made to compare the
thermographic image of damage with the actual damage in a composite
laminate. Whitcomb [19] used temperature profiles recorded on
thermographs during fatigue of boron-epoxy specimens as boundary
conditions in a two-dimensional finite element heat transfer program
to define the region of the heat source and thereby determine the
damage zone in the composite. The investigation was successful in
mapping the damaged area of material. The results also pointed out
that thermography is an integrated field measurement in that it
measures the collective heat produced by several damage mechanisms
without having to measure the contribution of each mechanism separately. This is a very desirable feature in composite materials
where the damage mechanisms are diverse and complex.

Jones [20] solved the heat transfer equation using finite difference techniques by generating heat at a known internal source and
computing isotherms on the surface of a composite plate. The calculated patterns were then compared with thermographs obtained from
composite panels with known defects using vibro-thermography.

Figure 8 shows a diagram of the intentionally included flaws and
their location in a $[(0,90)_2]_s$ graphite epoxy laminate. The mylar
squares are approximately one-eighth inch square, the teflon cloth is
one-half inch square, the diamond shaped mylar is about three-eights
inch across and the grease spots are approximately one-quarter inch
in diameter. Vibrothermograms of the graphite epoxy plate show the
grease spots and the teflon cloth very clearly. Isotherms resulting
from the solution of the heat transfer equation for these two cases
agree very well with the experimental thermograms. None of the mylar
pieces in the specimen were detected thermographically. Further
investigation revealed that the mylar bonds to the epoxy reasonably
well during curing to reduce the effect of the discontinuity at the
interface and reduce the heat generated by surfaces rubbing or

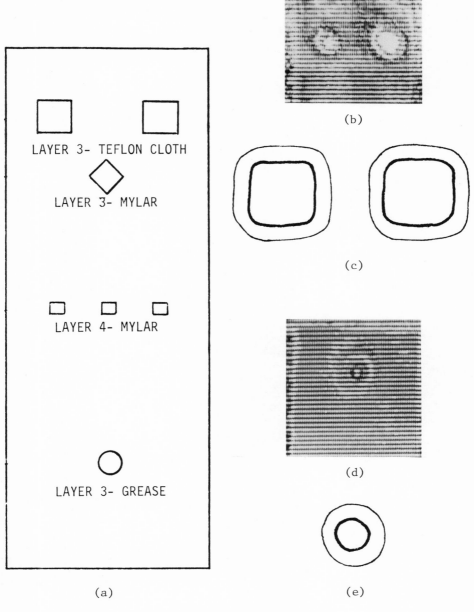

Fig. 8. (a) Intentionally included flaws in a $[(0,90)_2]_s$ graphite epoxy laminate.
(b) Thermograph and
(c) computed isotherms of teflon squares.
(d) Thermograph and
(e) computed isotherms of grease spot.

clapping together during vibration. Similarly the constraints imposed by bonded material along the boundary of the diagonally delaminated surfaces of the $[\pm45,0]_s$ specimen described previously could reduce the heat generated by the defect so that it was not detected thermographically.

CLOSURE

    A recent NDE technique, video-thermography, has been described from several viewpoints. The basic, infrared radiation phenomenon which is used to delineate the surface thermal gradients has been reviewed. A discussion of the mechanics of possible mechanisms responsible for the generation of heat from mechanical energy input has been presented. Finally, some experimental observations made with the technique on several different materials have been detailed.

    While still in the developmental stage, vibro-thermography appears to offer a wide and varied potential for nondestructive damage detection both for laboratory and field use. A unique aspect of video-thermography is that it is dependent on the stress state existing around a region of damage in a material. The mechanical energy input to the specimen is preferentially transformed to heat by the stress fields existing in the damaged region. It is likely that these stress fields are related to the mechanical strength, stiffness, and/or life of the material. Hence, the thermographic techniques described, which couple mechanical and thermal energy, offer the potential of directly monitoring the integrated damage stress state and predicting the mechanical response. Of course, additional analysis and experimental verification are required if this is to be realized. Finally, it is pointed out that the video-thermography technique is also distinctive in that it is a totally non-contact method. No interaction at all is required with the specimen to monitor the thermal gradients.

ACKNOWLEDGEMENT

    The authors wish to thank the following sponsoring agencies who, through the grants and contracts specified, have given their support to the work described herein: Army Research Office, Grant DAAG-29-79-G-0037; Air Force Materials Laboratory, Contract F33617-77-C-5044; National Science Foundation, Grant ENG 76-80213; and National Aeronautics and Space Administraton, Grant NSG-1364. Also, with much gratitude, we thank our secretary, Mrs. Phyllis Schmidt for her patient and diligent typing of this and other manuscripts.

REFERENCES

1. R. Vanzetti, Practical Applications of Infrared Techniques, (Wiley-Interscience Publication, John Wiley & Sons, New York, 1972).

2. E. G. Henneke, II and T. S. Jones, Detection of Damage in Composite Materials by Vibrothermography, "Nondestructive Evaluation and Flaw Criticality for Composite Materials," STP 696, Amer. Soc. for Testing and Materials, Phila. (1979).

3. R. E. Engelhardt, and W. A. Hewgley, "Thermal and Infrared Testing", Nondestructive Testing, A Survey, NASA-SP-5113, (National Aeronautics and Space Administration, Washington, D.C., 1973, pp. 119-140.

4. K. L. Reifsnider, E. G. Henneke, II, and W. W. Stinchcomb, "Defect Property Relationships in Composite Laminates," AFML-TR-76-81, Final Report, (June 1979).

5. F. A. McClintock and A. S. Argon, "Mechanical Behavior of Materials," Addison Wesley, pp. 471-486 (1966).

6. C. Zener, "Elasticity and Anelasticity of Metals," Univ. of Chicago Press (1948).

7. B. J. Lazan, "Damping Properties of Materials, Members and Composites," in Abramson, Liebowitz, Crowley and Junasz, eds., Applied Mechanics Surveys, Spartan Books, Washington, D.C., pp. 703-715 (1966).

8. B. J. Lazan, Damping Materials and Members in Structural Mechanics, Pergamon, New York, (1968).

9. R. Plunkett, "Mechanical Damping," Treatise on Analytical Chemistry, Kolthoff, Elving and Stross, eds., Part III, Vol. 4, pp. 487-516 (1977).

10. M. A. Biot, "Thermoelasticity and Irreversible Thermodynamics," J. Applied Physics, Vol 27, No. 3, pp. 240-253 (1956).

11. A. S. Nowick, Internal Friction in Metals, Prog. in Metal Physics 4, pp. 1-70 (1953).

12. W. P. Mason, "Piezoelectric Crystals and Their Application to Ultrasonics," D. Van Nostrand Co., Inc., New York (1950).

13. K. L. Reifsnider and W. W. Stinchcomb, Investigation of Dynamic Heat Emission Patterns in Mechanical and Chemical Systems, Proceedings of 2nd Biennial Infrared Information Exchange, AGA Corp, St. Louis, MO, pp. 45-58 (1974).

14. E. G. Henneke, II, K. L. Reifsnider and W. W. Stinchcomb, Thermography - An NDI Method for Damage Detection, Journal of Metals 31, pp. 1115 (1979).

15. James C. Hsieh, "Temperature Distribution Around a Propagating Crack," Doctoral Dissertation, College of Engineering, Virginia Polytechnic Institute and State University, (April 1977).

16.   L. C. Thomas, Fundamentals of Heat Transfer, Prentice-Hall,
      Inc., New Jersey, 1980.
17.   Yeung, P. C., "Investigation of Constraint Effects on Flaw
      Growth in Composite Laminates," Ph.D. Dissertation,
      Virginia Polytechnic Institute and State University,
      Blacksburg, VA 1979.
18.   Gibbins, M. N., "Investigation of the Fatigue Response
      of Composite Laminates with Internal Flaws," M.S. Thesis,
      Virginia Polytechnic Institute and State University,
      Blacksburg, VA 1979.
19.   Whitcomb, J. D., "Thermographic Measurement of Fatigue
      Damage," Composite Materials : Testing and Design
      (Fifth Conference), ASTM STP 674, S.W. Tsai, Ed., ASTM,
      1979, pp. 502-516.
20.   Jones, T. S., "Thermographic Detection of Damaged Regions
      in Fiber-Reinforced Composite Materials," M.S. Thesis,
      Virginia Polytechnic Institute and State University,
      Blacksburg, VA 1977.

DYNAMIC PHOTOELASTICITY AS AN AID TO SIZING

SURFACE CRACKS BY FREQUENCY ANALYSIS

A. Singh
C.P. Burger
L.W. Schmerr
L.W. Zachary

Ames Laboratory, USDOE, Engineering Research Institute
and Department of Engineering Science and Mechanics
Iowa State University
Ames, Iowa   50011

INTRODUCTION

This paper describes a method for sizing surface cracks that have been modeled as machined slots.[1] Several techniques have been used in the past to size surface cracks. Most use compressional (P) and shear (S) waves. Less attention has been given to using Rayleigh (R) waves or surface waves for sizing cracks. Since the energy of Rayleigh waves is confined to a layer of material near the surface, these waves have a great potential for sizing surface cracks. These are exactly the cracks that are hardest to characterize with traditional ultrasonic techniques because they lie in the "near-field" region of the transducers.

Dynamic photoelasticity was chosen to study the overall wave behavior and the mode conversions of a Rayleigh wave as it interacts with narrow slots cut from the edges of a two-dimensional plate model. This technique gives a full-field visualization of the stresses produced by an elastic wave traveling in a solid. The interaction between a Rayleigh wave and a slot could be observed from a sequence of pictures taken with a high-speed Cranz-Schardin camera. Rather than using the familiar amplitude technique, the authors have concentrated on the frequency analysis of the transmitted Rayleigh wave. The transmitted wave was chosen because it has interacted fully with the slot and should, therefore, contain the most information on the characteristics of the defect. The fact that the depth of the Rayleigh wave is a function of the wavelength was found to be the important property that can be used to find the depth of surface-breaking flaws. Fast Fourier frequency analysis of the surface wave showed that the high-frequency components of the input wave were selectively filtered by the slot. The results suggest that the wavelength corresponding to the highest frequency present in the transmitted wave will relate to the depth of the slot.

During recent years, investigators started to recognize the potential value of Rayleigh waves for characterizing surface and near-surface defects. This wave has its energy confined to a depth of approximately two times its wavelength so that it is not disturbed by internal or back-surface reflections. Reinhardt and Dally[2] used photoelastic visualization to study the interaction of Rayleigh waves with surface flaws. They found the variation of transmission and reflection coefficients for slots with depths of up to half of the Rayleigh wavelength. Bond[3] used a numerical technique to study the interaction of Rayleigh waves with slots. He indicated that the finite-difference modeling is a powerful technique for getting a quantitative understanding of the interaction and resulting scattered pulses from flaws with configurations that render them analytically intractable.

Work on using the Rayleigh-wave timing methods has predominantly been done in the United Kingdom. If the depth of the surface wave (about 2 λ) is smaller than the crack depth, then the near-surface part of the Rayleigh wave incident at the crack opening is forced to travel around the crack. The delayed time of arrival at a second receiving transducer can then be used to monitor the crack depth.

Silk[4] and Hall[5] both used this technique to size surface cracks by placing a transmitter and a receiver on opposite sides of a crack. Hall noted that three signals were received; he called the first the diffracted shear from the crack tip, the second the mode-converted shear wave from the Rayleigh wave, and the last the Rayleigh wave. In this study we found that the first and the second sets of signals noted by Hall are a mixture of two waves: the Rayleigh wave which forms at the crack tip from subsurface particle motion of that portion of the wave that is deeper than the crack, and the mode-converted and diffracted waves. Hall used the timing of the last signal, i.e., the Rayleigh wave that has traveled around the crack, to successfully size fatigue cracks in rails. He further clarified the sources of the three signals by using a photoelastic visualization technique in glass. The major drawback of the method is that for small cracks, the three signals merge together. Any timing method that relies on separating them is, therefore, unsuitable.

Silk[6] used a single surface-wave probe to size cracks. He used the timing of the direct reflections from the crack opening and the crack tip to find the depth of the crack.

Both the single-probe and the two-probe techniques are limited to deep cracks so that the various signals do not superimpose. It is possible to improve this technique by using very short pulses so that it becomes easier to differentiate various signals in the time domain.

Until recently, ultrasonic frequency analysis was mainly used to characterize internal flaws. Morgan[7] analyzed the reflected signal from a slot milled in aluminum and found certain modulations in the frequency spectrum. Bond[8] used a broadband (0.5-6 MHz) longitudinal wave transmitter, and placed it below the slot on the opposite face of the plate. A receiver placed on the side of the slot detected a Rayleigh wave. The peak frequency of this mode-converted Rayleigh wave was plotted for each slot. The peak frequency (f) of the Rayleigh wave was inversely proportional to the depth of the slot squared (slot depth α $1/f^2$).

The technique was used on slots 0.33 mm wide and from 0.5 to 3 mm deep in a duralumin cylinder.

The use of spectral analysis to determine the depth of surface cracks has tremendous potential for research because there are several different modes of converted and scattered signals that should be investigated to find how they correlate to crack depth. There is, as yet, no theoretical analysis available on the interaction of a wave with a slot. Experimental and numerical studies are needed to fully develop this method.

Achenbach[9] discussed the relative displacements for Rayleigh waves. The displacements are localized in a thin surface layer which has a thickness approximately equal to two times the wavelength. However, at a depth greater than 1.2 λ the amplitude of the motion is so small that it is often ignored and was not discernible within the resolution of the photoelastic system used in this research.

For photoelastic visualization, the use of a birefringent model material whose acoustic and elastic properties are as near as possible to those of the actual prototype material is recommended, i.e., steel or aluminum. For this requirement to be satisfied, the preferred model materials should be quartz or glass. The wave speeds in these materials are so high and the wavelengths generated by typical ultrasonic transducers so small that in order to "freeze" the wave motion extremely short exposure times are required (about 25-50 ns). Photoelastic visualization in glass and quartz has been studied by Hall[5] for British Rail. His results have been mostly qualitative.

In the work reported here, the model material chosen was Homalite-100. It has the drawback that its acoustic and elastic properties differ considerably from the actual test material. However, its high stress birefringence allows better visualization of the secondary waves. The wave speeds are much slower than in glass, and the wavelengths from the surface explosions that were used are longer so that exposure times of around 1 μs are sufficient.

A multiple-spark camera of the Cranz-Schardin type was used to get a sequence of pictures of the waves at different times. This technique was used by Burger et al.[10] and Riley et al.[11] to study the wave behavior in layered media. The camera can operate at a framing rate that can be varied in discrete steps from 67,000 frames/ s to 810,000 frames/s. The effective exposure time is around 0.6 μs.

DYNAMIC PHOTOELASTICITY

Dynamic photoelasticity provides full-field two-dimensional information on the propagation of elastic waves in sheets of bire-fringent material. Such information is valuable for the insight that it provides in the ways in which waves interact with a slot.

The major drawback of current photoelastic techniques is that in order to provide sufficient resolution of the stress field, highly birefringent model materials must be used. These materials are all polymeric, which means that they have very high attenuation for acoustic waves, low moduli, and low density, i.e., low acoustic impedance and wave speeds. The low modulus and high attenuation renders standard ultrasonic transducers incapable of inducing stress waves of sufficient magnitude to provide useful fringe information. In this research, a rapid-burning explosive (lead azide) was used to provide a sharp impulse wave.

Preparation of Model and Generation of Rayleigh Wave

The material chosen for the dynamic photoelasticity model was a 6.3-mm (1/4-in.) thick sheet of a polyester type material mar-keted by SGL Homalite under the trade name Homalite-100 (H-100). Its principal properties are listed in Table 1. Four plates with the geometry shown in Fig. 1 (but with four different slot depths) were used as indicated in Table 2.

The peak frequency of the input R-wave was 25 kHz, which corre-sponds to a wavelength of 44.4 mm. This peak wavelength is found from the Rayleigh wave velocity, which is tabulated in Table 1, and a plot of the frequency components of the signal, which will be given later. Since the input R-wave has a broadband frequency spec-trum (14 kHz to 110 kHz, approximately), the wavelengths of the

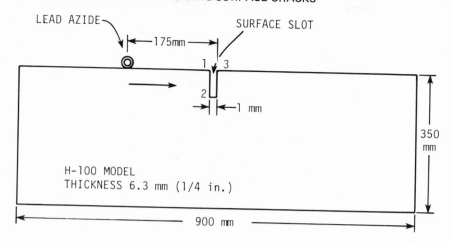

Fig. 1. Photoelastic model with a slot.

Fourier components vary from 10 mm to 80 mm about the peak value. The ratio of wavelength to the depth of slot is also shown in Table 2. A parametric study of the different effects of the four slots was used to find the ways in which changes in the frequency spectrum of the Rayleigh wave can be used to characterize the depths of the slots.

The R-wave was generated by exploding a small lead azide charge on the top edge of the plate. The explosive was packed into a tube, which was glued to the edge along a narrow line at a distance of approximately 175 mm from the slot. It was intended to approximate a line contact that would generate a cylindrical wave. The distance of 175 mm was chosen so that the reflected compressional or P-waves

Table 1. Dynamic Properties of Homolite-100

Velocities in H-100 Plate	
Longitudinal Wave	2.1×10^3 m/s (83,000 in./s)
Shear Wave	1.22×10^3 m/s (48,000 in./s)
Rayleigh Wave	1.11×10^3 m/s (43,500 in./s)
Elastic Modulus	
Tensile	2.4×10^9 N/m^2 (3.5×10^5 psi)
Compressive	4.5×10^9 N/m^2 (6.5×10^5 psi)
Specific Gravity	1.23

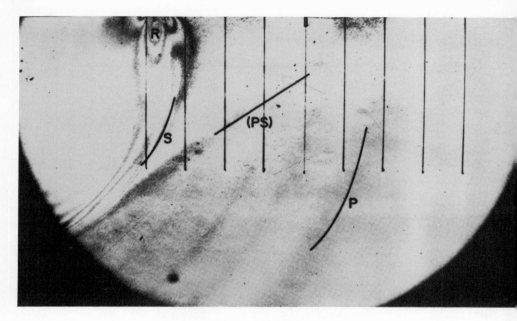

Fig. 2.   Input wavefronts.

from the edge and the bottom of the plate would not interfere with the surface or R-wave during the period of observation. The photo-elastic fringe pattern of the input wave before it interacts with the slot is shown in Fig. 2. The input R- and shear or S-waves are identified. This picture verifies the theoretical prediction of the waves generated by a suddenly applied line load.[12]

The following notation was devised to indicate the various waves. A capital letter indicates the type of wave, P for longi-tudinal wave, S for shear wave, PS for Von-Schmidt wave and R for Rayleigh wave. A subscript indicates the original wave from which the mode-converted wave (capital letter) was generated. A super-script identifies the point of origin on the model of reflected or mode-converted waves. Another superscript differentiates between a reflected (r) or transmitted (t) component.

In this manner the notation S_R^1 indicates a shear wave formed by mode conversion of the original R-wave at the first upper corner of the slot, Point 1 on Fig. 1. R_2^{r} will be the Rayleigh wave re-flected from the original R-wave at the slot tip (Point 2).

Interaction of a Rayleigh Wave with a Slot

The wave pattern that originates from the interaction of input

Table 2.   Dimensionless Ratios for the Plate Model

Model Number	Slot Depth (mm)	$\lambda^* = 44.4$ mm for R-wave			$\lambda = 10$ mm			$\lambda = 80$ mm		
		$\dfrac{\lambda^*}{h}$	$\dfrac{\lambda^*}{t}$	$\dfrac{\lambda^*}{b}$	$\dfrac{\lambda}{h}$	$\dfrac{\lambda}{t}$	$\dfrac{\lambda}{b}$	$\dfrac{\lambda}{h}$	$\dfrac{\lambda}{t}$	$\dfrac{\lambda}{b}$
1	2.8	15.9	44.4	7.0	3.6	10	1.6	28.8	80	12.5
2	6.9	6.4	44.4	7.0	1.4	10	1.6	11.2	80	12.5
3	9.9	4.5	44.4	7.0	1.0	10	1.6	8.0	80	12.5
4	12.9	3.5	44.4	7.0	0.8	10	1.6	6.4	80	12.5

λ^* = wavelength of peak frequency amplitude, i.e., at 25 kHz = 44.4 mm in Homalite-100

h = slot depth

t = slot width = 1 mm

b = plate thickness

waves with a thin slot in a plate is shown in Fig. 3. The important points at the slot are numbered 1, 2 and 3 as shown in the figure.

The photoelastic results show clearly that the pattern of mode conversions and the intensity of the various waves depend on the depth of the slot. The input S-wave is strong in the interior of the plate and reduces rapidly to zero at the surface. When it reaches the slot, most of its energy passes undisturbed past the tip. The portion of the shear wave that strikes the face of the slot is reflected and the portion that strikes the tip will diffract. For short slots, the diffracted shear is small since the incident S-wave (which is mainly a SV or head wave) is weak near the surface. Thus the incident shear wave striking the tip of the short slots is weak. As the slot depth increases, the intensity of the shear wave striking the tip will increase. The strength of the diffracted shear wave will increase correspondingly. It will again start decreasing as the effect of the larger diffraction angle becomes significant. The diffracted shear will be present at the surface, for instead of grazing the surface, it strikes at an angle.

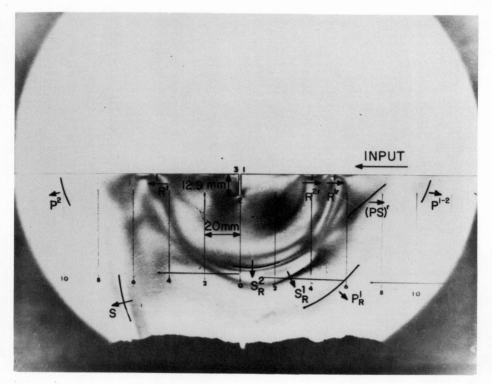

Fig. 3.   Waves generated by the input wave when it interacts with
          a 12.9-mm slot.

Also seen in Fig. 3 are two longitudinal waves P^2 and P^{1-2}.
It is not clear from which wave the two fronts P^2 and P^{1-2} are
generated.   The source for the former is Point 2.   For the latter,
the source may be Point 2 and/or the Face 1-2.

The interaction between the slot and the Rayleigh wave is the
most interesting.   It is known that the depth of the Rayleigh wave
is a function of the wavelength.   As the R-wave approaches the slot,
it strikes from the slot opening to the slot tip.   The particle
motion at the slot opening will contain all the Fourier components,
i.e., all the wavelengths of the input wave.   The particle motion at
the tip will be due mainly to the long wavelength sections of the
R-wave.

The interaction of the upper portion of the wave with the slot
opening (Point 1) is similar to the interaction of an R-wave with a
corner.   A R^{1r}-wave is reflected, a shear wave, S_R^1, is produced by
mode conversion from the R-wave at Point 1, and a transmitted Ray-

leigh wave turns around the corner and continues down the front face
of the slot to the tip. Here a portion of the R-wave mode converts
to a shear wave S_R^2, another portion is reflected back up the front
face as R^{2r}, and the remainder proceeds around the tip and up the
face of the slot.

The deeper particle motion of the R-wave will interact with
the slot differently. The energy distribution in the Rayleigh wave
as it interacts with the slot is sketched approximately in Fig. 4.
The deeper particle motion (corresponding to the long wavelengths)
goes under the slot and forms another Rayleigh wave, called here an
"undercut" R-wave. The deeper particle motion will also form a

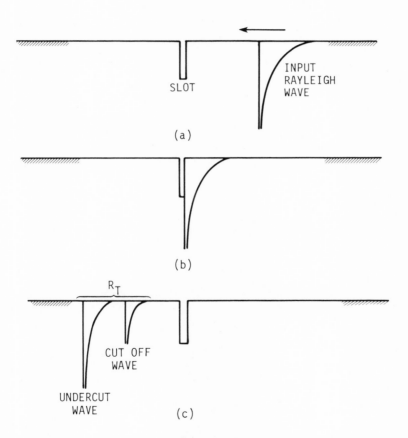

Fig. 4.   Formation of the cut-off and undercut Rayleigh waves. (The
          figure does not show other mode-converted and reflected
          waves.)

shear wave at the slot tip that will scatter. The transmitted signal R^t in Figs. 3, 4 and 5 is a composite of all R- and S-waves diffracted from the crack tip, including the shear wave that is scattered from the tip due to the incident shear wave. The shear wave
is present in the transmitted response at the surface, because unlike the input S-wave, it does not graze the surface but strikes at
an angle.

Photoelastic fringes represent the difference in the in-plane
principal stresses at each point of the dynamically stressed two-
dimensional model. The relationship between the fringe order N and
the stress difference is $(\sigma_1 - \sigma_2)\alpha N$. Since, at a free surface, one
of these stresses is zero, the fringe order at a free surface is
proportional to the tangential stress at the surface (N α σ_1, if
$\sigma_2 = 0$; N α σ_2, if $\sigma_1 = 0$). Thus the fringe order of a wave at a
free surface indicates the relative magnitude of the surface stresses
along that surface. These are plotted for the transmitted wave in
Fig. 5. This photograph is taken in "light field," i.e., the dark
areas represent half-order fringes. The white spot that is marked
R^t is a point of zero stress (zero-order fringe) where the Rayleigh

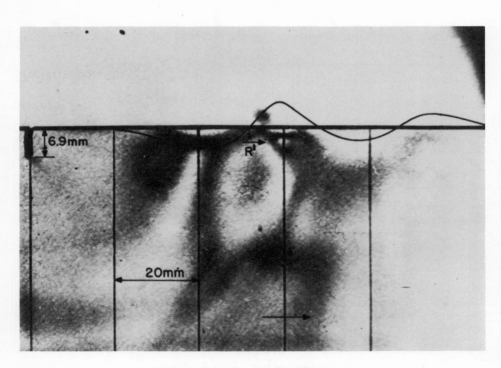

Fig. 5.   Transmitted wave.

wave changes from tension to compression. This point moves along
the surface from left to right at the Rayleigh wave velocity. The
plot of the fringe orders gives the shape of the transmitted disturbance. Since the slot depth affects the cut-off level, the shape
of the R^t-wave changes with the depth of the slot.

The results from the waves formed during the interaction with
the slot are in good agreement with Bond.[3] His results have all
waves except P^{1-2} and P^2. It is thus possible that these are not
generated by the R-wave, but by the input S-wave which is absent
in Bond's analysis. Thus, comparing the two results, it is found
that dynamic photoelasticity is indeed very useful in finding the
interaction of waves with discontinuities in a two-dimensional
medium.

Frequency Analysis of Transmitted Wave from Photoelasticity Data

The frequency of the transmitted wave is analyzed since it
contains the undercut (deeper part) R-wave, whose frequency depends
on the depth of the slot. Thus, it is possible that the frequency
spectrum of the transmitted wave has sufficient information to size
the crack.

In Fig. 6 the photoelastic fringe orders for different slots
are plotted along the top surface of the plate for the waves after
interaction with the slot. It represents the amplitude of the stress
wave along the surface. The position of the picture in the plate
field was chosen to show the "transmitted" R-waves, plus whatever
other waves happen to be clustered in the same region at the time of
observation.

The variation of the fringe order with distance along the surface (the spatial fringe distribution) was digitized by hand and
entered into a minicomputer system. A curve-fitting routine was
used to fit a cubic spline through these points. This wave was then
read automatically to give the adjusted fringe orders at equal in-
crements of 0.25 mm along the boundary. The fringe orders at these
incremental points define the wave very well. They are then taken
as the input for a Fast Fourier Transform (FFT) program. The FFT of
the signal contains both the real and the imaginary components of
frequency. Both components are added vectorially to obtain the magnitude of the frequency, or the frequency spectrum. This information
is plotted as the output of the program. Figure 7 shows the magnitude of the spatial frequency distribution for the transmitted waves
from each of the four slots. Since the transmitted wave is a mixture
of S- and R-waves which travel at different velocities, it is not
possible to convert these spatial plots to a single equivalent time
scale. There are significant differences in the frequency spectrums.

For the shortest slot (2.8 mm), the spectrum is quite similar to that of the input wave (Fig. 8), with some modulations. This is possibly due to resonances of the slot faces. As the slot depth is increased to 6.9 mm, the high-frequency content drops, while the low-frequency content remains strong.

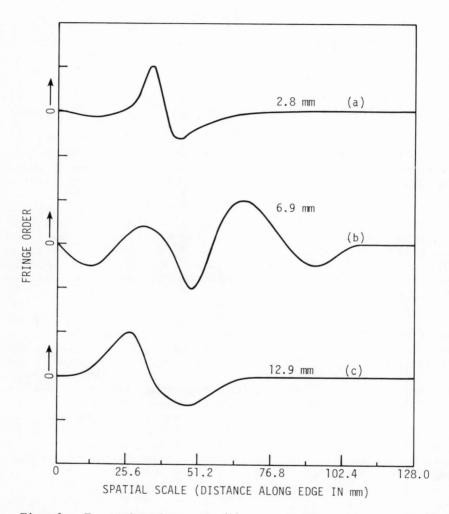

Fig. 6. Transmitted waves; (a) slot depth = 2.8 mm, scale
1 div = 3 fringes, (b) slot depth = 6.9 mm, scale
1 div = 1 fringe, (c) slot depth = 12.9 mm, scale
1 div = 1 fringe.

Since the depth of the R-wave depends on the wavelength, the portion of the wave that extends beyond the slot tip will pass and form the undercut R-wave and a mode-converted S-wave. The upper part of the R-wave that is "cut off" becomes very weak as it goes around the slot and loses significant energy into reflected R-waves and mode-converted waves. It is to be noted that this cut off part was the one that mainly contained the high-frequency components.

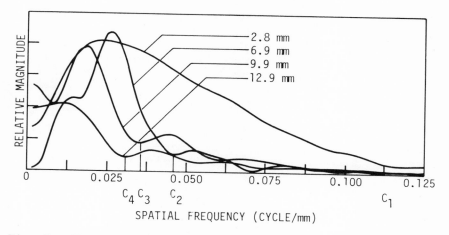

Fig. 7.   Frequency magnitude curves for four different slots.

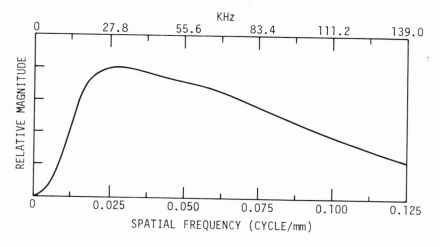

Fig. 8.   Frequency magnitude of the input R-wave for 2.8-mm slot.

The portion of this wave that is transmitted around the slot to appear as part of the transmitted R-wave on the far side of the slot
will have greatly reduced amplitude. Since this wave contains the
high-frequency components, the high-frequency portion of the transmitted spectrum will be weak. On the other hand, the undercut R-
wave contains the low-frequency content, which remains strong in
the transmitted wave. This argument holds true until the depth of
the slot becomes so great that there is no significant undercutting.
The shear wave that is formed by the deeper particle motion will
also be of low frequency as it is formed by low-frequency R-wave
components. With the increase in the slot depth, the input R-wave
is sliced at a lower level, which is the long wavelength region.

Thus, with the increase in slot depth, the frequency attenuation starts from the higher side and moves to the lower side. To
correlate this aspect quantitatively, it is necessary to relate a
particular point on the curve to the slot depth. One of the points
chosen was the point at which the frequency magnitude first goes to
a minimum. It was called the cut-off point because the frequency
components that are higher than the frequency corresponding to this
point were cut off by the slot. These points on the curves are indicated as C_1, C_2, C_3, and C_4, corresponding to the 2.8 mm, 6.9 mm,
9.9 mm and 12.9 mm slots. The frequencies and wavelengths corresponding to the cut-off points are tabulated in Table 3 and plotted
in Fig. 9.

There is, therefore, a direct relationship between slot depth
and the cut-off wavelength. This relationship was revealed by the
photoelastic data, but if it is to be practically useful, it should
be developed for characterization of surface cracks using Rayleigh-
wave transducers in ultrasonic testing.

Table 3. Frequency Data for Different Slots

Model Number	Slot (mm)	Cut-Off Frequency (cycles/mm)	Cut-Off Wavelength (mm)
1	2.8	0.112	8.9
2	6.9	0.046	21.7
3	9.9	0.035	28.6
4	12.9	0.031	32.3

CONCLUSIONS

The ability of dynamic photoelasticity to provide full-field views of elastic stress fields was used to provide an understanding of the ways in which the subsurface particle motions in Rayleigh waves are affected by a slot. As a consequence the property of the Rayleigh wave, which relates the wavelength to its depth below the surface, has been effectively used to find the depth of slots. The next step is to use conventional R-wave ultrasonic transducers on artificially machined slots or fatigue cracks to see how the slot depth relates to the cut-off wavelength. The transducers used should be broadband and the depth of the input R-wave should be greater than the slot depth so as to produce undercutting.

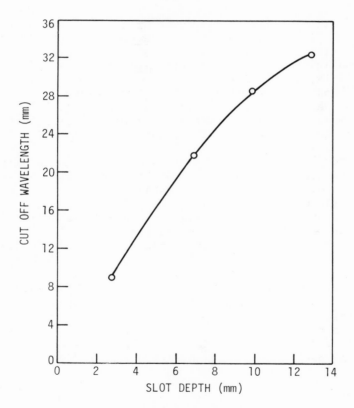

Fig. 9.   Slot depth vs cut-off wavelength.

ACKNOWLEDGMENTS

      This research was supported by the U.S. Department of Energy,
Contract No. W-7405-Eng-82, the Office of Basic Energy Sciences,
Division of Materials Sciences (AK-01-02), the Engineering Research
Institute and the Department of Engineering Science and Mechanics
at Iowa State University.

REFERENCES

1. A. Singh, "Crack Depth Determination by Ultrasonic Frequency
    Analysis Aided by Dynamic Photoelasticity," M.S. Thesis,
    Iowa State University, Ames, Iowa (1980).
2. H. W. Reinhardt and J. W. Dally, Some Characteristics of
    Rayleigh Wave Interaction with Surface Flaws, Materials
    Evaluation 28:213-220 (1970).
3. L. J. Bond, A Computer Model of the Interaction of Acoustic
    Surface Waves with Discontinuities, Ultrasonics 17:71-77
    (1979).
4. M. G. Silk, The Determination of Crack Penetration Using
    Ultrasonic Surface Waves, NDT International 9:290-297 (1976).
5. K. G. Hall, Crack Depth Measurement in Rail Steel by Rayleigh
    Visualization, Non-Destructive Testing 9:121-126 (1976).
6. M. G. Silk, Sizing Crack-like Defects by Ultrasonic Means,
    in: "Research Techniques in Non-Destructive Testing,"
    R. S. Sharpe, ed., Academic Press, London (1977).
7. L. L. Morgon, The Spectroscopic Determination of Surface
    Topography using Acoustic Surface Waves, Acustica 30:222-228
    (1974).
8. L. J. Bond, "Finite Difference Methods Applied to Ultrasonic
    Non-destructive Testing," Proc. of the meeting on ARPA/AFML
    Review of Progress in Quantitative NDE, La Jolla, CA, July
    1979 (To be published).
9. J. D. Achenbach, "Wave Propagation in Elastic Solids," North-
    Holland Publishing Company, Amsterdam (1973).
10. C. P. Burger and W. F. Riley, Effect of Impendence Mismatch
    on the Strength of Waves in Layered Solids, Experimental
    Mechanics 14:129-137 (1974).
11. W. F. Riley and J. W. Dally, A Photoelastic Analysis of Stress
    Wave Propagation in a Layered Model, Geophysics 31:881-899
    (1966).
12. M. J. Forrestal, L. E. Fugelso, G. L. Meidhardt and R. A.
    Felder, "Response of a Half Space to Transient Loads," Proc.
    of Engineering Mechanics Division Speciality Conference ASCE
    (1966).

ULTRASONIC AND X-RAY RADIOGRAPHIC INSPECTION AND CHARACTERIZATION

OF DEFECTS IN Gr/Aℓ COMPOSITES

T. Romano, L. Raymond, and L. Davis

Nevada Engineering and Technology Corp. (NETCO)
Long Beach, CA 90806

ABSTRACT

Ultrasonics and X-ray radiographic nondestructive inspection techniques have been evaluated and compared with the ultimate goal of developing quality assurance procedures for use in the manufacture of basic structural shapes of Graphite/Aluminum Metal Matrix Composites. To achieve this end, Gr/Aℓ wire, basic panel, structural shapes and stringer stiffened panels were nondestructively inspected, tested, and the defects microstructurally characterized. Flaw indications observed in X-ray radiographic and ultrasonic inspections were interpreted by metallographic examination. The Defect-Property Sensitivity Analysis (DPSA) program is a continuing effort directed at a more rigorous definition of test variables. (Refs. 1 through 4).

INTRODUCTION

Metal Matrix Composites (MMC) are finding their way into an increasing number of applications. As the number of applications increases so does the need for a reliable technique of NDE. This is needed to insure the quality of materials used and the integrity of the product during various stages of manufacture and also the final product.

NDE methods are being investigated by a number of agencies with the goal of achieving reliable NDE methods and the establishment of accept/reject criteria for materials and final products. In order to accomplish this task a knowledge of defect types evolving from manufacture and their detectability is essential. Further, it is essential to know the strengths and weakness of the NDE methods being employed.   (Refs. 6 through 8).

Research, at this time, has not demonstrated any one NDE method as being suitable for the entire task of inspection of MMC. X-ray radiography and ultrasonics are widely used methods of NDE and, therefore, are prime candidates for investigation of their usefulness in inspection of MMC.

Experimental Procedure

In order to establish the background required to achieve the ultimate objectives of the DPSA program, a three layer, 0.100 in. thick test plate composed of T50/201 graphite aluminum with various known defects was manufactured as a control sample. The intentionally introduced defects were:

1. Crossed wires in center layer.
2. Extra .004" thick surface foil.
3. 1/4 in. and 3/4 in. hole in center layer.
4. Stainless steel foil.
5. Various sizes and depths of flat bottom holes.
6. Debonded surface foil.
7. Unwetted fibers.
8. 1/4 in. diameter lead tape on surface (used for calibration).

X-ray inspection was accomplished by standard inspection techniques with the goal of achieving the best resolution. Ultrasonic C-scan was accomplished varying transducer figuring and diameter, and the pulser/receiver gain and gate level. This established a basis for inspection of plates with unknown defects.

A number of Gr/Al plates were ultrasonically inspected by using an immersion system and a through transmission technique. Through transmission was accomplished by reflecting the sound beam off a glass reflector plate located one inch from the plate (Fig. 1). The reflected signal is gated at 20% signal amplitude. This makes it possible to monitor the attenuation of the sound beam traveling through the plate (Fig. 2). A 10 MHz, 1/2 inch focused transducer was used to produce the sound beam and was driven by a Krautkramer USIP 11 pulse echo scope (Ref. 5).

A C-scan recorder was used to produce a plan view of the discontinuities that are recorded as white areas on a dark gray background (Fig. 3). These white areas are produced when the signal amplitude is attentuated below the 20% reference level of the gate.

Calibration was accomplished by using a 1/4 inch diameter lead tape with a .003 inch thickness. The gain level was adjusted to produce a .025-inch diameter indication when the area of the tape

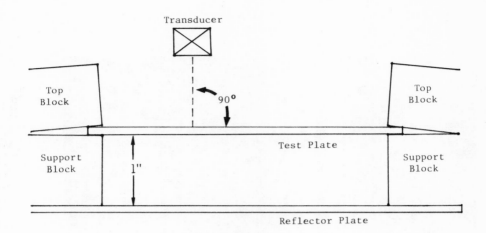

Fig. 1.   Test plate set-up.

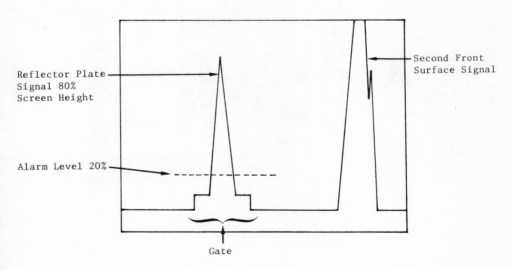

Fig. 2.   Gate adjustments on screen display

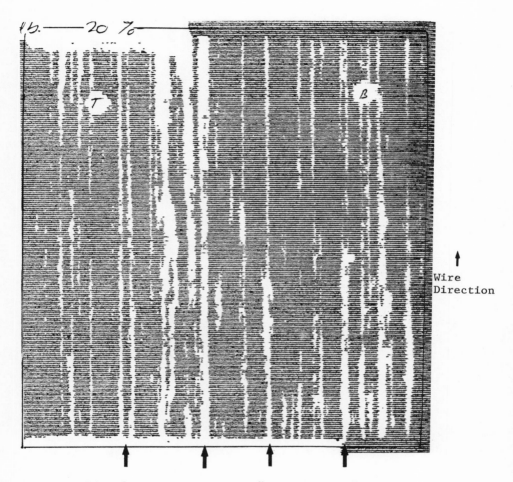

Fig. 3.   C-scan showing "longitudinal" ultrasonic
          indications of a graphite-aluminum plate.

is scanned (Ref. 4).

RESULTS

Classification of NDE Indications

Examination of the C-scans and X-radiograph revealed that the various indications could be divided into four groups. The size and shape of the indications were used as guidelines for determining the following classifications of indications (Table I).

Table I - Grouping of NDE Indications

Indication	Description
Longitudinal	Linear indications parallel to fiber direction.
Area	Local indication with average diameter of the area being greater than .10 of an inch.
Spot	Local indication with average diameter of the area being less than .10 inch.
Ring	Local indication generally outlining defective surface area.

Ultrasonic Defect and Indication Correlation

In an effort to correlate defect types with indications, microstructural examination was performed on selected areas of the plate and specimens were sectioned and prepared. The plate selected was a two layer, unidirectional Gr-Aℓ plate containing T50 fiber. Correlation of the microstructure with ultrasonic indications produces the following grouping of defect types with NDE indications.

Table II - Causes of Ultrasonic Indications

Indication	Defect Type
Longitudinal	- Unwetted longitudinal fibers - Longitudinal cracks - Different fibers (Figure 4) - Voids due to incomplete consolidation of wires (Fig. 6)
Area	- Delamination including unbonded surface foil (Fig. 7)
Spot	- Voids (Fig. 5) - Excess aluminum (Fig. 4) - Inclusions
Ring	- Missing surface foil

X-ray Defect and Indication Correlation

Similar correlation of microstructure with X-ray radiography allowed the following grouping of defects with X-ray indications.

Table III - Causes of X-ray Indications*

NDE Indication	Defect Type
Longitudinal	- Unwetted fiber - Longitudinal cracks - Different fiber (Fig. 4) - Voids - incomplete consolidation (Fig. 6) - Columnation (fiber) variations
Area	- Missing foil - Foil variations (thickness)
Spot, white	- Voids (Fig. 5)
Spot, dark	- Excess aluminum (Fig. 4) - Inclusions

* Interpretation accomplished using positive print of X-ray negative.

DISCUSSION

Comparison of the ultrasonic and X-ray defect classifications reveals that debonds, including surface foil debonds, are not

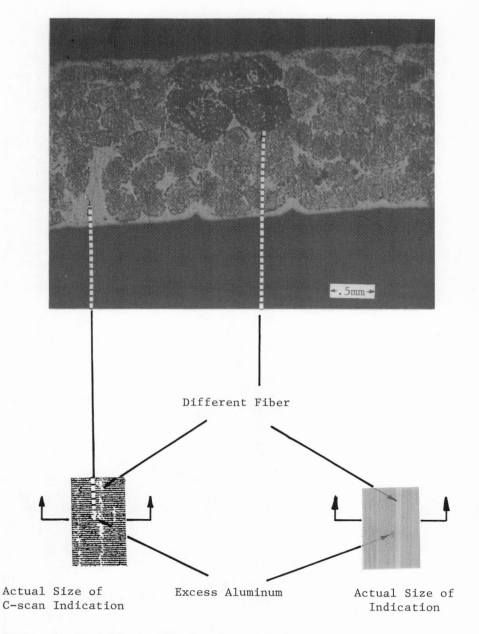

Figure 4.   Metallographic Section of Defective Area Caused by the
            Presence of Different Fibers and Excess Aluminum
            Correlated to C-Scan and X-Ray Indications.

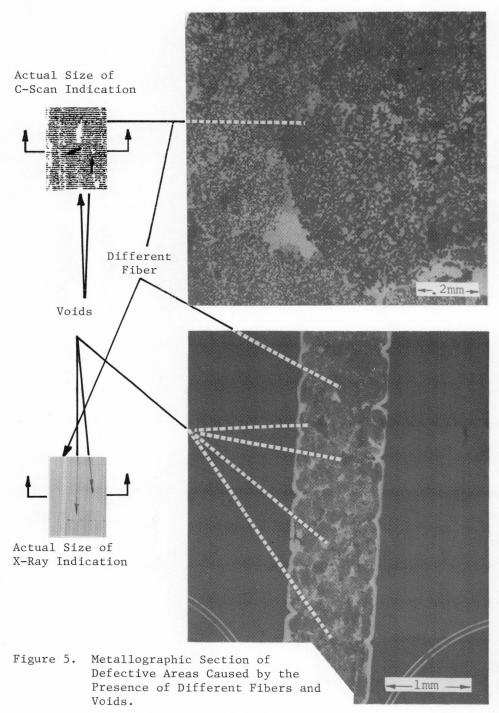

Figure 5.   Metallographic Section of
            Defective Areas Caused by the
            Presence of Different Fibers and
            Voids.

Actual Size of
X-Ray Indication

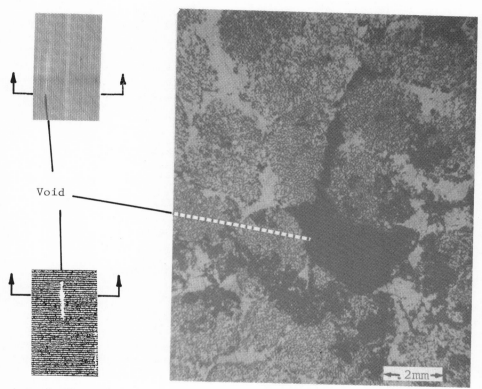

Void

Actual Size of
C-Scan Indication

Figure 6. Metallographic Section of Defective Area Caused by the
Presence of a Longitudinal Void Correlated to C-Scan
and X-Ray Indications.

visible with X-ray radiography but are easily detected by the C-scan (Figure 7). On the other hand, fiber columnation is clearly seen in the X-ray print but not in the C-scan.

Foil variations listed under area indication in the X-ray classification has not been fully verified but observations indicate that thickness variations produce area indications which look like dark gray patches on the plate similar to an aerial view of patchy fog. Fiber columnation is visible in these areas. The darker areas would indicate denser material (thicker foil) if these observations prove to be correct.

Voids are listed under Longitudinal and White Spot Indications. This is due to the fact that microstructural studies revealed voids due to incomplete consolidation which gave a longitudinal indication and voids which were not longitudinal but produced a white spot indication.

It can be seen at this point that these NDE methods have their advantages and disadvantages. However, these disadvantages can be overcome by using the C-scan and X-ray jointly. Positive identification of voids, delaminations, and missing surface foil can be made.

X-ray allows for better defect identification because it allows the fiber orientation to be viewed. As an example of this, black spot indications can be separated into two types. One type is viewed as being between the wires (Figure 8) and in some instances, it is separating the fiber at that point. This would indicate possible excess aluminum between the wires. The other type of black spot indication is viewed as superimposed (Figure 8) and would, therefore, be surface foil related. This could be an inclusion or excess aluminum. Both of these defects would appear as spot indications on the C-scan with no visible means of making any distinctions.

The ability to see the fiber is also helpful in detecting variations in spacing of the wires and wire orientation. The ability to detect these wire related characteristics decreases as the layers of wire increase.

In cases where the C-scan indicates an area indication, debonding, it is also possible that the large area indication could be "masking out" other smaller internal indications. Using the X-ray and the C-scan together will allow detection of additional anomalies such as voids, excess aluminum, etc.

SUMMARY

In an effort to summarize the results of this research effort,

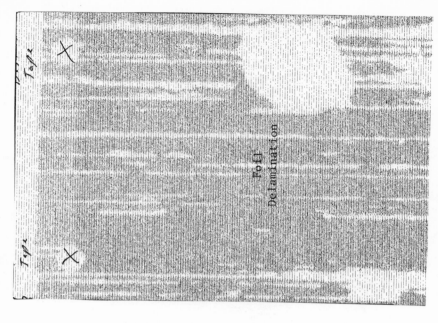

Figure 7. C-Scan and X-Ray Showing Delamination Not Visible in X-Ray.

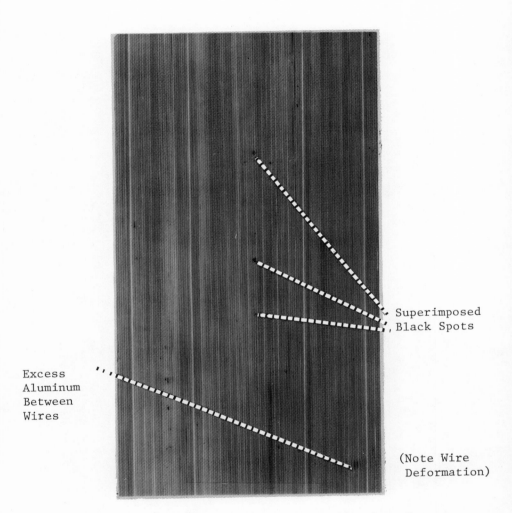

Figure 8.   X-Ray Showing Black Spot Indications Between
            Wires and Superimposed.

defect types were grouped with the three components of metal matrix
composites. Table IV is a grouping of defects which are fiber
matrix or foil related.

Table IV

Component	Defect Type
Fiber	− Unwetted − Twisted − Broken − Misaligned − Different − Crossed
Matrix	− Voids − Inclusions − Cracks (longitudinal and transverse) − Incomplete consolidation
Foil	− Missing − Extras − Different − Delamination − Thickness Variation − Holes, tapered, top, bottom, side

CONCLUSIONS

In Gr/Aℓ composite plates, the introduction of a variety of
defects is extremely useful in establishing the settings for the
optimum resolution of ultrasonic and X-ray radiographic techniques.
Metallographic characterization is a necessary tool to understand
the source of the indication. Four different groups and 16 different types of defects were identified.

Ultrasonics and X-ray radiography must be used in combination
in order to locate and classify the type of defects, but neither
NDE method can separate the active from the passive type defects
relative to their end uses; i.e., orientation of defect relative
to the loads. In addition, neither method has the potential of
predicting the fracture strength of a composite. The latter two
issues must be resolved before NDE of metal matrix composites
becomes a useful tool. This is why the Defect Property Sensitivity
Analysis (DPSA) program is a continuing effort. It becomes impera-
tive that an understanding be developed of the sensitivity of the

mechanical properties to the size, shape and location of a fabrication defect relative to the end-use of the product.

> Longitudinal ultrasonic and X-ray indications are generally unwetted fibers, different fibers, longitudinal cracks, or voids due to incomplete consolidation.

> Area indications are basically foil related. The foils are generally used to provide transverse strength.

> Spot indications consist primarily of voids, excess aluminum, and inclusions.

> Ultrasonic C-scan does not show fiber columnation but easily detects debonds.

> X-ray radiography cannot detect debonds but gives visible representation of fiber columnation.

> At this time, insufficient data exists for a precise evaluation of the sensitivity of mechanical strength to various types of defects until more research studies are completed.

REFERENCES

1. Davis, L. W. and Sullivan, P. G., "Exploratory Development of Graphite-Aluminum Composites," NETCO Final Technical Report to Naval Sea Systems Command under Contract N00024-76-C-5334, October 1977.

2. Pfeifer, W. H. and Young, R. C., "The Development of Graphite Aluminum Structural Hardware for Launch Vehicle Application," NETCO Final Technical Report to Naval Surface Weapons Center under Contract N60921-76-C-0219, March 1, 1978.

3. Pfeifer, W. H., "The Development of an Advanced Metal Matrix Composites Program Plan for DoD Applications," NETCO Quarterly Progress Report No. 4 to Naval Surface Weapons Center, November 1, 1977.

4. Raymond, L. and Romano, T., "Review of and Future Trends in NDE of Metal Matrix Composites," Paper presented at 4th Annual Conference of Composites and Advanced Ceramic Materials, Cocoa Beach, Florida, January 1980.

5. "Production of Graphite-Aluminum Sheet, Plate and Structural Shapes with Inspection, Quality Control and Characterization," DWA Composite Specialties, Inc. Final Report of Naval Ordnance Station under Contract N00197-78-C-0150, September 12, 1979.

6. Harrigan, W. C., Jr. and Hudson, S. P., "Inspection of Graphite-
    Aluminum Composites by Means of Holography," Prepared for
    Space and Missile Systems Organization, Air Force Systems
    Command under Contract No. F040701-76-C-0077, June 21, 1977.

7. Anderson, C. W. and Blessing, G. V., "Detection of Material
    Defects in Graphite-Reinforced Aluminum," NSWCTR 79-222,
    September 1979.

8. Chaskelis, H. H., "Study of the Problems Associated with the
    Ultrasonic Inspection of Graphite-Aluminum Composites," Paper
    presented at the 4th Annual Conference of Composites and
    Advanced Ceramic Materials, Cocoa Beach, Florida, January 1980.

EVALUATION OF SENSITIVITY OF ULTRASONIC DETECTION

OF DELAMINATIONS IN GRAPHITE/EPOXY LAMINATES

S. W. Schramm
I. M. Daniel
W. G. Hamilton

Materials Technology Division
IIT Research Institute
Chicago, Illinois 60616

ABSTRACT

A study was conducted to evaluate, improve and optimize ultrasonic techniques for the detection of tight delaminations in composite laminates. Four graphite/epoxy panels were investigated using
matched pairs of compression and shear wave ultrasonic transducers
of three frequencies (1, 2.25 and 5MHz). Pulse echo and through
transmission modes were used and A-scans, C-scans and frequency
spectra were analyzed. It was found that the higher frequency (5MHz)
compression transducer operated in the pulse-echo mode gave the best
results in most cases. Shear wave transducers were found to be cum-
bersome to use and unsuitable for the cases studied.

INTRODUCTION

Composite materials offer many attractive features in the design
of aircraft, naval and automotive vehicles. In order to employ composites in design with more confidence it is necessary to be able to
assess their integrity and strength by nondestructive means. One of
the most widely used methods of nondestructive evaluation is the ultrasonic method with its many variations and techniques.

Various kinds of flaws can be introduced into a fiber reinforced
composite laminate during its fabrication and processing. Some of
these are: contaminants, voids, unpolymerized resin, resin-rich and
resin-starved areas, ply gaps, inclusions, delaminations, crazing and
surface scratches. Tight edge delaminations may also develop due to

drilling and cutting after laminate cure. Such tight delaminations
are generally not detectable by visual inspection, but can be discovered by nondestructive inspection (NDI) techniques. Many initial
flaws tend to grow under load and service conditions, spreading
damage and starting new flaws.

Several techniques exist for nondestructive detection of flaws
in materials.[1] Of interest here is the ultrasonic method as applied
to composite materials. The method is often used in the aircraft
industry, where many of the techniques for panel inspection have
evolved.[2-4] Examples of its application in a research environment
are inspection of composite flat plate specimens for fabrication
nonuniformities, and intentionally placed artificial flaws.[5-9]

The work described in this paper was part of a larger program
"Sensitivity of Ultrasonic Detection of Delaminations and Disbonds
in Composite Laminates and Joints", the objective of which was to
evaluate and improve the sensitivity of ultrasonic detection of tight
delaminations in composite panels and disbonds in composite/metal
joints. The study was conducted for the David W. Taylor Naval Research and Development Center.

Methods and principles for ultrasonic inspection and flaw detection in materials are detailed in the literature.[1,10,11] Basically a high frequency sound (1 to 25MHz) is emitted in periodic
bursts from an ultrasonic transducer through a coupling medium into
the specimen material being inspected. The resulting attenuated
pulse emerging from the specimen is picked up by a receiving transducer; the information from it is electronically processed; and the
data is displayed for the evaluation of presence, size and location
of flaws.

The preferred method of flaw identification is by comparison
of pulse information obtained from known flawed and unflawed standard specimens. Appropriate standards, especially in laminated composites, are not always easy to produce and sometimes are not feasible. Alternate pulse processing/identification techniques for ultrasonic inspection, such as frequency and phase shift analysis and
adaptive learning techniques, are being developed.[5,12-15]

The program was conducted using ultrasonic transducers of 1,
2.25 and 5MHz frequencies operating in either the pulse echo or
through transmission mode. Two types of ultrasonic waves were used,
compressive and shear, so the specimen could be inspected as thoroughly as possible.

The experimental data generated during the project was recorded
by using an X-Y plotter for C-scans, oscilloscope photographs for
A-scans and frequency distribution photographs for RF spectrum analysis. The photographs are used primarily to supplement C-scan records

EXPERIMENTAL PROCEDURE

Specimens

Four graphite/epoxy laminate specimens were inspected. These specimens were grouped into three primary types

Type 1. Two 0.64 cm (0.25 in.) thick graphite/epoxy panels of unknown quality. (Fig. 1a)

Type 2. A 1.27 cm (0.5 in.) thick graphite/epoxy laminate with artificial flaws of relatively low sonic transparency at known locations. (Fig. 1b)

Type 3. A 1.27 cm (0.5 in.) thick graphite/epoxy laminate with artificial flaws of relatively high sonic transparency at known locations. (Fig. 1c)

Ultrasonic Inspection System

Transducers. Two different types of transducers were selected for inspection of the specimens; they were compressive and shear wave transducers. A matched set of transducers of each wave type, and of each of three frequencies, 1, 2.25 and 5MHz, was selected. A total of twelve transducers were used in this program.

The requirements for determining the compressive wave transducer selection were: a standoff distance of at least 5.08 cm (2 in.) to clear hardware associated with specimen mounting; high damping and broad band width to provide resolving power for detection of subsurface flaws; and a small soundbeam cross-section for good definition of flaw boundaries in the plane of the specimen.

The compressive wave transducers are immersion type focused transducers 2.54 cm (1 in.) in diameter. Their nominal focal length in water is 6.4 cm (2.5 in.) The focal spot size was found to be 1.3 mm (0.050 in.) which makes it possible to outline the boundary of a flaw in the plane of the specimen to within ± 0.64 mm (± 0.025 in.)

The requirements for determining the shear wave transducer selection were: a normal incidence angle into the specimen; and high damping and broad band width also to provide resolving power for the detection of subsurface flaws.

The shear wave transducers are contact type transducers 1.27 cm (0.5 in.) in diameter. They must be used with a shear wave couplant between the transducer and specimen to ensure transmission of the ultrasonic wave into the specimen. A transducer holder, shown in

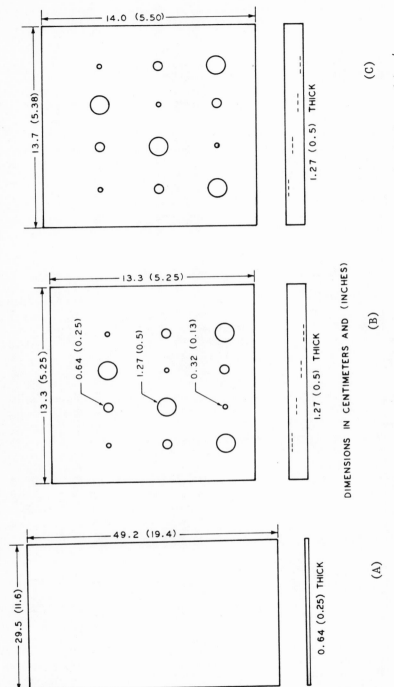

DIMENSIONS IN CENTIMETERS AND (INCHES)

Figure 1. Graphite/epoxy laminate specimens. (A) Type 1: 0.64 cm (0.25 in.) thick graphite/epoxy panel of unknown quality. (B) Type 2: 1.27 cm (0.5 in.) thick graphite/epoxy laminate with artificial flaws of relatively low sonic transparency at known locations. (C) Type 3: 1.27 cm (0.5 in.) thick graphite/epoxy laminate with artificial flaws of relatively high sonic transparency at known locations.

Fig. 2 was used to maintain maximum contact between the transducer and specimen via the shear wave couplant.

Ultrasonic Pulser-Receiver. Figure 3 shows the overall setup of the ultrasonic scanning and recording system used in this program. The ultrasonic transducer system is operated by a Model 5052 UA pulser-receiver manufactured by Panametrics, Inc. It is capable of driving the transducers in either the pulse-echo or in the through-transmission mode.[1,10] The pulser-receiver is a broadband device which can readily operate transducers in the range of 1MHz to 20MHz. It has independent countrols for input energy, signal repetition rate, high pass filter gain and attenuation all of which control the final transducer signal. It provides a time domain output for re-viewing the R.F. signal on an oscilloscope or spectrum analyzer. It also has a gated peak output detector which can operate one axis of an X-Y recorder pen for plotting the ultrasonic information contained in the variation of the peak output signal.

Support Equipment. In addition to the transducers and pulser-receiver four pieces of equipment are required to get a complete picture of the specimen integrity: they are an X-Y recorder, an oscilloscope, R.F. frequency spectrum analyzer and an oscilloscope camera.

The X-Y plotter has a control sensitivity range from 0.25 mV/cm to 5 V/cm. The recorded C-scan can be smaller, equal to or larger than the actual specimen up to a size of 25 x 38 cm (10 x 15 in.) The X-Y recorder can operate in either the pen-lift or analog mode. In the pen-lift mode the output from the peak detector is channeled through an electronic alarm unit into the pen-lift control of the recorder. The alarm unit is set to trigger the pen-lift whenever the amplitude of the peack detector falls below a prescribed level, indicating a flaw, otherwise the pen stays in contact with the paper tracing a line. This gives detailed flaw locations in the plane of the specimen with definite outlines of the flaw boundaries. The analog mode supplements this information by giving a continuous record of the amplitude of the gated pulse. It is obtained by scanning the specimen with the alarm unit bypassed. The horizontal and vertical motion of the pen is affected by information fed to the X-Y recorder by displacement transducers used on the scanning drive mechanism. The scanning drive mechanism, which will be described later, like the X-Y recorder, are used in conjunction with the compression wave immersion transducers only.

The oscilloscope used was a 50MHz, dual channel portable oscil-loscope. The oscilloscope is fed with information processed by the pulser-receiver. A trace is generated off the marked R.F. terminal of the pulser-receiver with a W-shaped pulse representing the front face reflection and a smaller M-shaped pulse representing the back face reflection. A second trace is a voltage/time representation

Figure 2.   Shear wave transducer in through transmission mode without specimen.

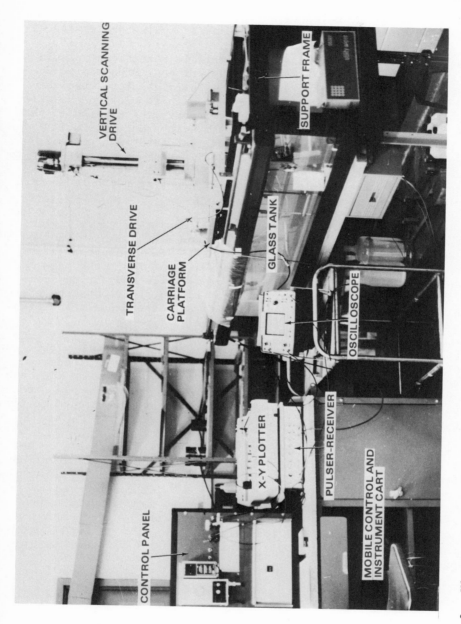

Figure 3. Ultrasonic scanning and recording system for nondestructive flaw detection in composite laminates.

of the gated portion of the pulse reflected from the back face. This
trace is generated off the gated R.F. terminal of the puler-receiver.
The gate window may be moved to any position between the front and
back faces by adjusting the range knobs on the pulser-receiver.

The spectrum analyzer used was a Variable Persistence display
type with High Resolution IF section and 1kHz to 110MHz tuning section. With these modules the analyzer measures absolute frequency
and amplitude, has frequency and amplitude high resolution capability,
has high sensitivity and provides means of observing signals at the
gated R.F. terminal on the pulser-receiver. The range value in the
tuning section of the analyzer was set at the operating frequency of
the transducer being used. A nominal bandwidth of 300kHz was used to
get a relatively broad window. The photographs of the oscilloscope
and spectrum analyzer signals were made with a screen mounted camera
with Polaroid pack film back.

Scanning System. The scanning system consists of an all glass
immersion tank 45.7 cm (18 in.) wide, 45.7 cm (18 in.) high and 183
cm (72 in.) long resting on a support table. The transparent glass
tank affords distortion-free visual observation through the tank
sides during aligning of specimens and transducers and during scanning. A steel frame resting on the support table surrounds the glass
tank which is accurately positioned relative to the frame. The frame
supports the mechanical system for scanning the specimens.

The mechanical system has a scanning carriage platform with
three precision screw drives which move the ultrasonic transducer
along three mutually perpendicular axes. It is guided along the tank
length on steel shafts attached to the support frame. The platform
is elevated above the tank and can be traversed and positioned at any
tank location.

A motorized screw drive is used for vertical motion and the vertical scanning arm holding the ultrasonic transducer at its lower end.
The drive is capable of moving the arm in up and down scanning strokes
of any adjustable length up to 35.6 cm (14 in.) at rates up to 127
cm/min (50 in./min.) The strokes are monitored by the cable-type
electromechanical position transducer attached to the vertical drive.
A manually operated horizontal transverse screw drive is used for locating or focusing the ultrasonic transducer each time anew specimen
is scanned.

General Procedures

The objective of the program was to determine the "optimal scanning method" for each of the three types of specimens. The term "optimal scanning method" implies the best combination of transducer
type and frequency plus operational mode such that as many as

possible of the known artificial flaws could be correctly detected and characterized.

The program was divided into the following tasks:

1) Select transducers with characteristics most likely to fulfill the objective of this program.

2) Generate the most representative "flaw map" of each specimen using combinations of operational modes and transducer frequencies.

3) Develop criteria for determining optimum combination of operational mode and transducer frequency for generating the best "flaw map" for each type of specimen.

4) Draw conclusions and make observations on correlations between specimen types and their "optimal scanning method."

To accomplish the tasks as efficiently as possible the following procedure was followed. The first step was to start working with a specimen type which had the physical characteristics closest to those types of specimens with which we had the most experience. The Type 1 0.64 cm (0.25 in.) thick graphite/epoxy specimens were first analyzed using the 5MHz compressive wave immersion transducer in the pulse echo mode. The 5MHz frequency was selected because in previous applications it was found to be the most suited to this type of specimen and its pulse could be more readily interpreted. The pulse-echo mode was selected becasue when analyzed on the oscilloscope display, the echo pulse can be used to locate approximately the position of a flaw through the thickness of the specimen.

The second step was to compare the results from step one with the specification sheet of the specimen. Then an improvement was sought by changing the control settings on the pulser-receiver in a progressive manner until an "optimal representation was achieved.

Once a specimen type was "optimized" the pulse-receiver settings were recorded and used as the starting point for the analysis of the successive specimen types. All the specimen types were analyzed in this manner before either the operational mode or transducer frequency were changed. The specimens were analyzed in sequence from Type 1 (thinnest, 0.64 cm thick graphite) to Type 3 (thickest, 1.27 cm graphite.)

An identical procedure was followed with the normal incidence shear wave transducers with the exception that a solid steel specimen was used as a standard against which laminated specimen results could be compared.

The data generated during the program was recorded as either a
C-scan, an A-scan, an R.F. frequency spectrum photograph or a combination of the three. When compressive wave transducers were used
all three recording methods could be used. Since the shear wave
transducers required point per point application only A-scans and
spectrum photographs could be used.

The C-scan is a two-dimensional representation of material integrity as determined by a compressive wave tranducer used in conjunction with the ultrasonic scanning and recording system (Fig. 3).
The flaw indications in the pen-lift scan depend entirely on the
trigger level set on the alarm unit. This level must be based on
known standard flaws. In this case the natural standard for gapless
delaminations was used to set the trigger level. The A-scan is an
amplitude-time display for a given point on the specimen and can indicate the approximate location of a flaw through the thickness of
the specimen.

The R.F. frequency spectrum photographs can be used to supplement the information contained in the A-scans. The frequency spectrum is a representation of the "signature" of the gated portion of
the R.F. pulse as it passes through a specimen. The presence of delaminations, artifical flaws, or adhesive interfaces causes a filtering effect of the ultrasonic wave which shows as a frequency spectrum
shift, hence, a characteristic "signature" is generated. At this
point it is difficult to make any general conclusions about the R.F.
signature because of the uniqueness of the specimen types investigated

RESULTS

Results on the optimization of ultrasonic detection are presented
here for each of the three specimen types investigated.

Type 1 Specimens

These were two 0.64 cm (0.25 in.) thick graphite/epoxy panels of
unknown quality. Figure 4 shows a full scale C-scan of one section
of the specimen. The 5MHz compressive wave immersion transducer operated in the pulse echo mode was determined to be the best for the
following reasons. The surface irregularities of the specimen could
be easily defined because of the focused nature of the compressive
transducer.

It was discovered immediately that shear wave transducers could
be discounted for further considerations as an "optimal method."
Every specimen type was inspected using the shear wave transducers,
but, in no case were they found to be competitive with the compressive wave transducers.

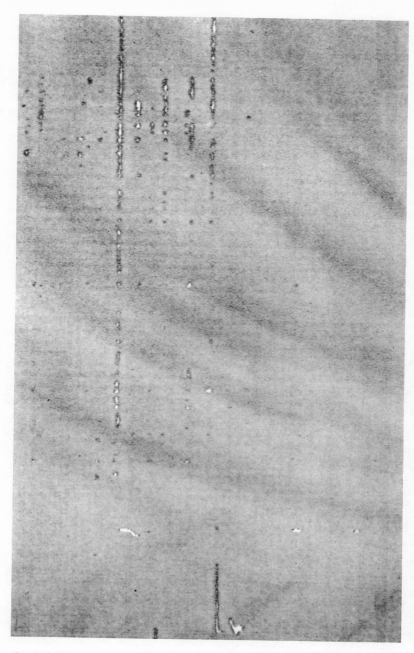

Figure 4. Full scale 5 MHz pulse-echo C-scan representation of section 2 of specimen Type 1.

Type 2 Specimen

    This was a 1.27 cm (0.5 in.) thick graphite/epoxy laminate with
artificial flaws of relatively low sonic transparency at known locations. Figure 5 shows the full scale C-scan of the plate. The 5MHz
compressive wave immersion transducer operated in the through-transmission mode was determined to be the best. All but the 0.32 cm
(0.13 in.) diameter artificial flaw located between plies 42 and 43
could be found. Reorientation of the specimen was attempted but
there were no conditions under which the flaw could be detected.

Type 3 Specimen

    This was a 1.27 cm (0.5 in.) thick graphite composite laminate
with artificial flaws or relatively high sonic transparency at known
locations. Figure 6 shows the full scale C-scan of the plate. The
5MHz compressive wave immersion transducer operated in a modified
pulse echo mode gave the best results. For this specimen the gate
window was positioned at the known interface of the two plies between which the artificial flaws were located. The movement of the
gate window away from the usual back face position is the reason for
the negative appearance of the scan. This is due, in part, to how
the trigger is set on the pen-lift alarm voltage control. The halos
which appear to outline the flaws are the areas of tight delaminations between the two plies caused by the presence of the flaws.

    Some general observations can be made by reviewing the optimal
C-scans for each specimen. First, even though unique specimen types
were inspected, all of them could be evaluated without the use of
additional special equipment or procedures not normally used in the
ultrasonic inspection of materials. This, coupled with the fact that
all three specimen types could be analyzed with one transducer frequency proves the flexibility and general applicability of the standard ultrasonic detection method. The only unusual technique tried
was the modified gate pulse echo on specimen Type 3, and this simply
required the adjustment of a single control on the pulser-receiver.

    The 5MHz compressive wave immersion transducer was the best for
the majority of specimens because of the material being inspected.
The shear wave transducers were found to experience extensive wave
energy dissipation by the laminated structure of the graphite/epoxy.
This occurs because with the shear wave the particles oscillate normally to its direction of travel, i.e., parallel to the ply interfaces. Increased wave energy is just dissipated over a wider contact
area. Also, the 5MHz transducer was found to be the best in most
cases because its high frequency and shorter wavelength make it more
sensitive in detecting and defining flaws, especially tight delaminations and those of relatively high sonic transparency. The 5MHz
signal is more prone to being scattered by tight delaminations,

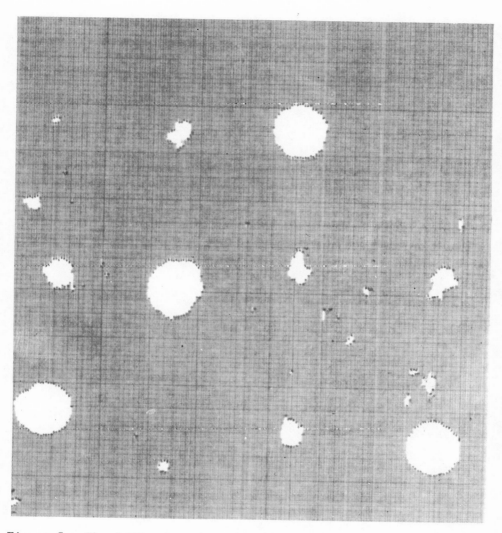

Figure 5. The 5 MHz through transmission C-scan representation of
specimen Type 2.

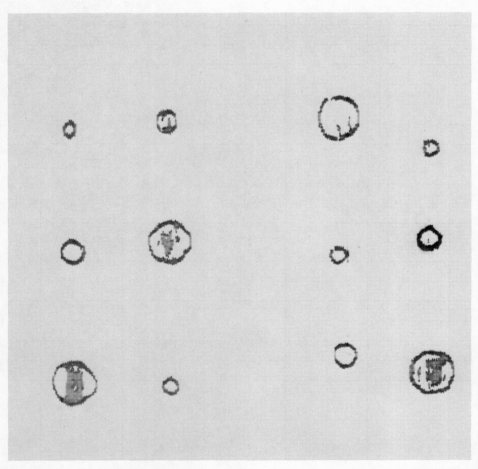

Figure 6.   The 5 MHz modified pulse-echo C-scan representation of specimen Type 3.

thereby showing a flawed area on the C-scan, because the wavelength is closer to the order of magnitude of the flaw size than the other frequencies.

SUMMARY AND CONCLUSIONS

The objective of this program was to evaluate and improve the sensitivity of the ultrasonic detection of tight delaminations and disbonds in composite panels and joints.

The program was performed using an existing ultrasonic detection and recording system with two transducer types, shear and compressive wave, in two operation modes, pulse echo and through transmission, at three frequencies, 1, 2.25 and 5MHz. Each of the three specimen types was exhaustively inspected using all possible combinations of operational variables until a unique set of optimal scanning settings were found and a flaw map was generated.

It was found that, due to their operational characteristics, normal incidence shear wave contact transducers are not applicable for use with composite laminates. The 5MHz compressive wave immersion transducer was clearly the best for the three specimen types. Of the two operation modes the pulse echo mode was best suited for detecting tight delaminations in the composite specimens. Whenever both pulse echo and through transmission techniques generate C-scans of comparable quality, the pulse echo mode is preferred as it is easier to use because of specimen fixturing problems associated with using the automatic scanner in the through transmission mode. A-scans and R.F. spectrum photographs provide excellent support data for flaw characterization, but neither C-scan or A-scan alone are totally adequate to define the position of a flaw within a specimen. The ability to determine the presence of artificial flaws is more dependent upon the location of the flaw rather than whether the flaw is of high or low sonic transparency. Finally, exotic techniques or equipment do not have to be used to get consistent results for specimens covering a wide range of physical specifications.

1. R. C. McMaster, "Nondestructive Testing Handbook," The Ronald Press, New York (1963).
2. D. Mool, R. Stephenson, Ultrasonic Inspection of a Boron/Epoxy-Aluminum Composite Panel, Materials Evaluation, 29:7:159-164 (1971).
3. J. F. Moore, Development of Ultrasonic Testing Techniques for Saturn Honeycomb Heat Shields, Materials Evaluation 25:2:25-32 (1967).
4. J. B. Whiteside, I. M. Daniel, and R. E. Rowlands, "The Behavior of Advanced Filamentary Composite Plates with Cutouts," Technical Report AFFDL-TR-73-48, (1973).
5. J. L. Rose, J. M. Carson, and D. J. Leidel, Ultrasonic Procedures for Inspecting Composite Tubes, "Analysis of the

Test Methods for High Modulus Fibers and Composites,"
ASTM STP 521, pp 311-325, (1973).

6. D. E. Pettit, Residual Strength Degradation for Advanced
    Composites, Interim Tech. Qrtly. Report, LR 28360-5,
    Lockheed-California Co., for AFFDL, (5 August 1978).

7. G. P. Sendeckyj, G. E. Maddux, and N. A. Tracy, Compari-
    son of Holographic, Radiographic, and Ultrasonic Tech-
    niques for Damage Detection in Composite Materials,
    ICCM/2 Proc. of 1978 Int'l Conf. on Composite Mat'ls,
    Toronto, Canada, (16-20 April 1978)

8. S. V. Kulkarni, R. B. Pipes, R. L. Ramkumar, and W. R.
    Scott, The Analytical, Experimental and Nondestructive
    Evaluation of the Criticality of an Interlaminar Defect
    in a Composite Laminate, ICCM/2 Proc. of 1978 Int'l
    Conf. on Composite Mat'ls, Toronto, Canada, (16-20
    April 1978).

9. W. C. Harrigan, Jr., Ultrasonic Inspection of Graphite/
    Aluminum Composites, ICCM/2 Proc. of 1978 Int'l Conf.
    on Composite Mat'ls, Toronto, Canada, (16-20 April 1978).

10. J. Krautkrämer, and H. Krautkrämer, "Ultrasonic Testing of
    Materials, Springer-Verlag New York, Inc., New York,
    (1969).

11. J. R. Frederic, "Ultrasonic Engineering, John Wiley & Sons,
    Inc., New York, (1965).

12. O. R. Gericke, Ultrasonic Spectroscopy, in: "Research
    Techniques in Nondestructive Testing," R. S. Sharpe,
    ed., Academic Press, New York, (1970) pp 31-61.

13. F. H. Change, R. A. Kline, and J. R. Bell, Ultrasonic Eval-
    uation of Adhesive Bond Strength Using Spectroscopic
    Techniques, General Dynamics, Fort Worth Dov., Proc.
    ARPA/ARML Review of Progress in Quantitative NDE,
    Scripps Inst. of Oceanography, La Jolla, Calif., (18
    July 1978).

14. "Interdisciplinary Program for Quantitative Flaw Detec-
    tion," Rockwell Int'l Science Center, Contract No.
    F33615-74-C-5180, for ARPA/AFML, (Sept. 1975)

15. A. N. Mucciardi, "Adaptive Nonlinear Signal Processing for
    Characterization of Ultrasonic NDE Waveforms," Task 2,
    Measurement of Subsurface Fatigue Crack Size, Technical
    Report AFML-TR-76-44, (April 1976).

INFLUENCE OF INITIAL DEFECT DISTRIBUTION ON

THE LIFE OF THE COLD LEG PIPING SYSTEM

B. N. Leis, M. E. Mayfield,
T. P. Forte, and R. J. Eiber

Battelle
Columbus Laboratories
Columbus, Ohio 43201

ABSTRACT

There are two basic design philosophies to avoid fatigue fracture presently contained in the ASME Boiler and Pressure Vessel
Code. One, safe life, provides for fatigue resistance by avoiding
the initiation of cracks, whereas the second, damage tolerance,
presumes the existance of cracks and defects and guards against
their unstable growth. This paper examines the utility of these
approaches in the context of the sensitivity of piping system life
to the character of the presumed initial defect distribution.
The importance of defect detectability and growth monitoring is
illustrated and the significance of parameters like defect size
and shape to damage tolerant approaches discussed, in terms of
piping system life. Results of the paper are interpreted in light
of relevant NDI capabilities and requirements.

INTRODUCTION

There are two basic design philosophies to avoid fatigue fracture presently contained in the ASME Boiler and Pressure Vessel
Code. One, "safe-life", is embodied in Section III and detailed
by example in Code Case 1592. The other, linear elastic fracture
mechanics (LEFM) based "damage tolerance", has been more recently
introduced into the code and is embodied in Section XI. It is
noteworthy that the first of these philosophies seeks to avoid
the initiation and growth of cracks to a critical length by sizing
components based on the fatigue resistance of polished small-diameter
specimens. It is, therefore, predominately a philosophy which

325

seeks to avoid the initiation of cracks. In strong contrast, the
damage tolerant philosophy seeks to avoid the crack growth of presumed existing defects to a critical size. Between these two philosophies lies a gray region. It is associated with the growth
of cracks that may be said to have initiated, based on the small-
diameter specimen data. However, such cracks are quite small render-
ing their detection and growth monitoring, necessary within the
damage tolerant philosophy, quite difficult.

This paper is concerned with the detectability of cracks in
the just noted gray region. The paper numerically examines the
sensitivity of the life of reactor cold leg piping to the initial
defect distribution, to establish the significance of reliable
NDI in piping vessel design. The results of this study are then
interpreted with regard to (1) NDI inspection capabilities and
requirements, and (2) the viability of either safe-life or LEFM
damage tolerant philosophies as being self-sufficient for Boiler
and Pressure Vessel design codes. These two aspects are considered
within the framework of LEFM. Consequently, the paper first examines
LEFM in applications to piping analysis, detailing in particular
the material properties, defect character, and the assumptions,
simplifications and limitations of the analysis. Next, the results
of the numerical synthesis are presented in the format of a sensitivity study of piping life as a function of defect character.
These results are then considered in the context of current nondestructive inspection (NDI) system sensitivity, defect detectability
(probability of detection/nondetection) and piping system inspection
schedule requirements. The paper concludes that defect detection
schemes beyond current NDI capabilities may be required to guard
against the development of large breaks in critical components.

APPROACH - LEFM ANALYSIS OF PIPING COMPONENTS

Basic Concepts - LEFM and Damage Tolerance

LEFM concepts are well documented in any number of books and
papers (e.g., see Reference 1). Our purpose here is, therefore,
not to introduce and detail the concepts but to outline the salient
features and inherent assumptions.

The essence of LEFM is the stress intensity parameter, K,[2]
which is a measure of the crack driving force in terms of the far
field stress(es) and crack dimension(s). In a simple functional
form K is given for Mode I cracking by:

$$K = \beta S\sqrt{\pi a} \quad . \tag{1}$$

In this equation β is a geometric function which may also depend
on the far field stress, S, and the semi-crack length, a.

Empirical observations over the last 20 years have shown that the range of stress intensity, ΔK, calculated from the range of cyclic stress, ΔS, is a unique damage parameter for fatigue crack growth per cycle, da/dN, under constant amplitude load cycling at a given stress ratio, R (K_{min}/K_{max}). It is generally accepted that, for confined flow at the crack tip, the growth rate is related to ΔK and R:

$$\frac{da}{dN} = f(\Delta K, R) \quad . \tag{2}$$

The functional form of such a relationship would also hold for variable amplitude cycling so long as the current plastic zone size is uniquely related to ΔK, in that LEFM requires such a unique relationship.

Given the form of Equation (2), one could estimate the period of crack growth, N_p, by rearranging it in the form

$$N_p = \int_{a_o}^{a_f} \frac{da}{f(\Delta K, R)} \tag{3}$$

where the limits on the integral are the initial, a_o, and final, a_f, crack sizes. Introducing Equation (1) into Equation (3) leads to

$$N_p = \int_{a_o}^{a_f} \frac{da}{f(\beta \Delta S \sqrt{\pi a}, R)} \quad . \tag{4}$$

The propagation period, therefore, depends on (1) the loading through S and R, (2) the geometry through β, (3) the material properties through the function f, and (4) the initial flaw character through a_o and β. The other parameter, a_f, depends on the failure criterion used, most of which embody some measure of the material toughness (e.g., K_{IC}).

Equation (4) permits one to estimate the development of defect size and shape. With the advent of LEFM and its application in the context of Equation (4) came the development of a new design philosophy – Damage Tolerance[3]. With its roots in the aerospace industry, it is especially suitable for dealing with higher strength materials which, because of their relatively low toughness, are very defect sensitive. A key aspect of this design philosophy is the setting of inspection intervals based on computed defect

growth rates. An equally key aspect is the associated inspection
of the structure. Such an inspection serves to provide physically
measured insurance of the structure's integrity. With favorable
experience, which shows computed rates compare with or over-estimate
actual growth, one gains confidence in predictions which indicate
that defects will not grow to a critical size during the next inspection interval.

The following sections of this paper examine and discuss computed growth rates and inspection intervals for piping products.
In contrast to many of the aerospace applications, this discussion
will focus on very ductile, lower strength, materials with usually
very high toughness. However, before this can be done the parameters
which enter into Equation (4) will be defined. Note that an indepth
description of these parameters can be found in Reference 4. Note
too that these parameters exist within the framework of LEFM analysis
of piping. Therefore, it is appropriate to first review critical
assumptions, simplifications and limitations.

Critical Assumptions, Simplifications and Limitations

A number of assumptions and simplifications have been made
in developing the results presented later in the sensitivity analysis.
Some are inherent in LEFM whereas others are a consequence of having
to synthesize otherwise unavailable data, and to adapt and implement
the analysis.
Key assumptions and simplifications include
- various loading components are combined via superposition
  to define a stress cycle which when it exceeds the yield
  stress of the material is pseudo-elastic,
- the strain equivalence principal is invoked to reduce
  this pseudo-elastic stress to a value compatible with
  material stress-strain response,
- through-the-wall stress gradients are considered as a
  uniform stress with a magnitude equal to the peak of
  the gradient,
- stress cycles for various transients are combined without
  regard to phase and plane of action,
- K solutions for dissimilar metal welds and nozzles developed from available results of experiments and analyses
  are valid: LEFM is applied in the analysis of the very
  ductile piping materials for which the confined flow
  assumption may be violated,
- stress redistribution does not occur as a consequence
  of crack growth,
- cracks grow in a self-similar fashion,
- ISI's fail to detect growing defects (cracks),
- fatigue is the dominant damage mechanism,
- crack damage accumulates linearly (sequence independent).

Other assumptions have been made as detailed along with their justification in Reference 4. All assumptions have been made with a bias towards conservatism, but an attempt has been made to retain realism. The major limitation of the analyses pertains to the applicability of LEFM to the very ductile piping products.

Inputs to the Analysis

Because the intent of this paper is to examine the sensitivity of life to the character of the initial defect distribution, the inputs are detailed only to the extent necessary for continuity.

Loads and Stresses. Basically there are three types of loadings imposed on a piping system, (1) pressure, (2) moment, and (3) thermal. Stresses due to pressure, moment, and thermal loadings were superposed on a worst case basis without regard to their sense or temporal base. In many cases, this superposition resulted in primary plus secondary stress values which exceeded the yield strength of the material. In such instances, the maximum stress was reduced by invoking the strain equivalence principle, the essence of which is employed in current fracture mechanics analysis procedures as detailed in Reference 5.

All transients postulated to occur during the life of a PWR, plus a stress attributed to steady-state vibration have been considered. The vibration is considered to be a consequence of a variety of loadings for example, pump vibration. The vibration stress range was assumed to be 1 ksi based on engineering judgement while the frequency of the loading was chosen as 1000 cpm, the running speed of the main coolant pumps.

Once the stresses were determined for each of the transients, they were arranged into a spectrum. The ordering of the stresses in the spectrum could not be based on service experience. Thus, the transients and associated stress ranges were arbitrarily distributed uniformly during the plant life.

With respect to fabrication and service induced mean stresses, high stress ratio data ($R \simeq 0.7$) were used in the crack growth rate analysis in an attempt to indirectly account for the presence of high mean stresses.

Materials Data. Material properties embodied in the analysis include: yield and ultimate strength; fracture toughness (K_{IC}, J_{IC}); and fatigue crack growth rate. All relevant properties are detailed in Reference 4. Of particular importance in the present paper are the fatigue crack growth rate data for carbon and stainless steels. Data used in this analysis for carbon steels are derived from published data for SA-533 Grade B.[6,7] The data used

for the stainless steels is based on published values for cast
stainless steel.[8]

The fatigue crack growth data for the carbon steel are shown
in Fig. 1 to illustrate such data. With respect to Fig. 1, it
is interesting to note the shape of the upper bound curve. At
low ΔK's, the data falls on or below the 'water' line presented
in Section XI of the Code. However, at approximately a ΔK of 10
ksi $\sqrt{in}$ the growth rate rises sharply above the ASME curve for
a water environment by about 1-1/2 orders of magnitude. As ΔK
continues to increase, the growth rate trend tends back toward
the ASME curve. Clearly, the Section XI line appears nonconservative
in this region, based on the data presented.

With regard to the analyses, the upper bound trend to high
stress ratio data has been used in an attempt to indirectly account
for scatter and fabrication/service induced positive mean stresses,
respectively.

<u>Stress Intensity Solution/Failure Criteria</u>. The solution
of Equation (4) requires knowledge of the function $f(\beta \Delta S \sqrt{\pi a}, R)$.
As noted above, the dependency on R has been indirectly included
by using high stress ratio data. Thus, one only needs a fracture
mechanics model that incorporates the component geometry, stress
range, crack size, and crack geometry. This model is a mapping
function that relates the crack driving force in the simple cracked
geometry used to develop material property data to that of a crack
in the structure, subjected to a specific set of loadings.

In dealing with the very ductile piping materials the fatigue
crack growth and residual strength analysis would ideally be based
on Elastic Plastic Fracture Mechanics (EPFM) rather than LEFM.
Unfortunately, however, EPFM has not been developed to the point
that it can be reliably applied to component geometries beyond
the simple laboratory specimens. Thus, the models were developed
in the framework of LEFM concepts for the fatigue crack growth
analysis. Both LEFM and 'plastic collaspe' concepts were utilized
in the context of the leak-before-break criteria (termed herein
failure criteria). The term 'plastic collaspe concepts' loosely
denotes both LEFM empirically corrected to include large-scale
yielding and the more rigorous EPFM formulations.

A fracture mechanics model was prepared for axial and circumferential surface and through-wall cracks in straight pipe and elbows*,
nozzle corner cracks, axial through-wall cracks at nozzles, and
cracks at dissimilar metal welds. The failure criteria were used

---

*Elbows were treated as straight pipe with the elbow stress distri-
 bution imposed.

to evaluate the final crack size, a_f, while the fracture mechanics models were used in the fatigue crack growth analysis. These models represent adaptations of published stress intensity factor solutions, the details of which may be found in Reference 4.

Defect Character. The final parameter that enters the analysis is the shape and size of the initial defect and its distribution within the pipe. Because the model is purely deterministic, a specific set of shapes and sizes must be chosen and then presumed to exist with specific orientations at critical areas in the structure. Such is a worst case analysis; yet it may be quite realistic in that the defect which causes problems is always the worst one, having the worst orientation with the highest stresses. Failures are due to defects which grow--the distribution of the benign, while interesting, has little bearing on the life of a structure which is limited by the growth of a single defect.

Consider now the size and distribution of defects within the pipe. Size must be based on the capabilities of NDI practices. The size required here is the largest defect size that might be missed rather than the smallest size that can be detected in view of the just noted facts. The shape of the defect also enters in the expressions for K. From a detectability standpoint, the crack shape and depth combine to establish the initial crack size a_o for a specific shape. Thus, to provide the necessary input to the analysis one must know the shape of flaws that might be found in piping and the largest flaw size that might be missed by current NDI methods.

A review of pertinent literature as discussed in Reference 4 revealed virtually no information that was relevant to the detection of fatigue cracks in reactor piping and very little information concerning the shape of fatigue cracks in piping. Therefore, it was necessary to estimate crack sizes that might be missed in an in-service inspection.

The types of materials used in the cold leg piping are (1) carbon steels, (2) wrought stainless steels, and (3) centrifically cast stainless steels. Based on comments from personnel involved in the inspection of piping, flaw sizes as a function of the wall thickness were established for each type. Specifically, for the carbon steels an initial crack depth of 25 percent of the wall thickness was chosen while an initial depth of 50 percent was chosen for the stainless steels (both wrought and cast forms).

The crack shapes chosen for the analysis were based on fatigue crack shapes found in failure investigations. The general shape is semi-elliptical with depth to surface length ratios (a/2c) between 0.1 and 0.5.

The orientation of initial defects, with respect to the pipe axis, was chosen as either longitudinal or circumferential, typically both being examined.

SENSITIVITY ANALYSIS OF PIPING LIFE
TO INITIAL DEFECT CHARACTER

In order to assess the sensitivity of piping life to defect character one must first establish the basis to define life.

Failure Criteria for Piping
Definitions of Life

In a general sense, there are several criteria which may serve to define failure or the end of the useful life of a piping system. These criteria include:
- The development of a through crack which is associated with leak development, from a part-through wall crack,
- The detection of a given leak rate after through crack development, and
- The development of a large break which severely curtails the pipe's function.

NDI plays a role in each of these, although not in the traditional sense for the leak rate criterion. With regard to the first, NDI techniques would have to reliably detect part-through wall cracks. For the second and third criteria, NDI would have to detect through-wall leaking cracks so that sensors, sensitive to leakage and humidity rather than cracks may be most appropriate. Of course, in all cases, the detection must come well before the crack growth becomes unstable to ensure the system can be safely shut down. And, because safety is best insured by early detection, NDI is relied upon for all criteria to detect part-through defects in both the preservice inspection (PSI) and the inservice inspection (ISI). It is the detection of part through defects that is focused on here.

Analysis Format

The following sections demonstrate the sensitivity of plant life to initial defect character in terms of each of the above criteria. Results are presented in graphical form on coordinates of crack length and cycles of load application, measured in terms of the number of times a history, presumed representative of one plants operating life (40 years) is applied. Because the transients are uniformly distributed in time in this history, results generated are only valid for intergral blocks of plant life. Plant life based on the first criterion is given as the terminal point, denoted by x, on plots of part-through crack length versus blocks of loading. Typical figures will be presented which reflect results developed for the three plants considered: Arkansas Nuclear (AN-1), St. Lucie (SL-1), and Farley (FL-1). Since the trends are comparable from plant to plant, actual data for only one plant will be presented and discussed in the remainder of this paper. As stated earlier, plant life based on either of the second two failure criteria, while developed, are not essential to the purpose of the paper and therefore are not considered.

Because several variables are examined in the study, more than one curve is presented in each figure. For this reason, the curves are number-keyed to a matrix of variables examined.

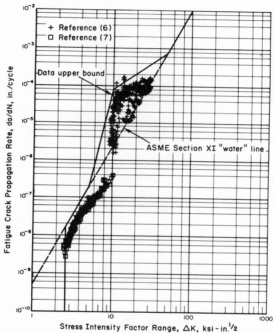

Fig. 1. Fatigue Crack Growth Rate Data for SA-533 Grade B--Conditions of High R, Laboratory Air, and Simulated PWR Environment

As part of the sensitivity analysis two initial flaw sizes will be examined. Further, several different shapes will be considered in both axial and circumferential orientations at different locations in the piping system. However, since the growth of a given defect is a consequence of the various loads imposed and the materials characteristics, the sensitivity of piping life to changes in defect character must be studied in the context of a given set of loadings and materials characteristics. Thus to put the sensitivity study into perspective one must first consider

the crack growth behavior in light of the loadings applied and
the fatigue resistance assumed in the analysis. The only significant change in the loadings from those normally considered in reactor
design is the inclusion of vibration. In regard to materials behavior, the data shown in Fig. 1 have been used in contrast to
the ASME Section XI curve.

Results and Discussion

Vibration and the Threshold. The impact of vibration on life
predictions and the dependence of this impact on the threshold,
ΔK_{th}, have been studied for each of the three plants as detailed
in Table 1. Note from the table that it relates specifically to
axial flaws with a/2c = 0.1. The sensitivity study has been done
for specific points in the piping system that represent the highest
stressed regions based on the analysis detailed in Reference 4.
The influence of vibration is evident by comparing curves 1 through
4 with 5 through 8, and 9 and 10 with 11 and 12. Since the trends
are comparable from plant to plant, only those for one plant are
presented in Fig. 2 and discussed here.

Table 1. Matrix for the Sensitivity Study on Vibration,
          Threshold, Growth Rate and Initial Flaw Size for
          Axial Cracks with a/2c = 0.1[a]

Curve[a] Number	ΔK_{th},[b] ksi√in.	Vibration Loads	da/dN[c]	a_i,[d] %t
1	Yes	Yes	B	25
2	Yes	Yes	B	10
3	Yes	Yes	A	25
4	Yes	Yes	A	10
5	Yes	No	B	25
6	Yes	No	B	10
7	Yes	No	A	25
8	Yes	No	A	10
9	No	Yes	B	10
10	No	Yes	A	10
11	No	No	B	10
12	No	No	A	10

[a]Results are reported in Fig. 2.
[b]ΔK_{th} = 2.6 ksi√in., carbon steels; ΔK_{th} = 2.4 ksi√in.,
 stainless steel.
[c]A - ASME line, B - composite data for carbon steel.
[d]Use 50 percent and 10 percent t for stainless steel.

With respect to the presence of vibration, consider first

curves 1 through 4 contrasted with curves 5 through 8. It is evident from these curves that, when a threshold, ΔK_{th}, of 2.6 ksi $\sqrt{in}$ is assumed, vibration gives rise to only nominal differences in piping life. On a relative basis its effect appears most severe for the larger initial flaw size, a trend attributed to the fact that more of the loading exceeds the threshold level. Vibration becomes a significant life-limiting-loading, when a threshold is not assumed to exist (or if it exists, its level is near zero). Under this condition, curves 9 and 10 indicate that the piping system life is near zero, a result which is apparently inconsistent with the service experience of PWR plants.

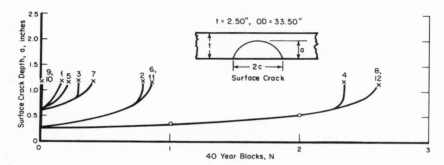

Fig. 2.   Typical Results of the Sensitivity Study on Vibration, Threshold, Growth Rate and Initial Flaw Size for Axial Cracks With a/2c = 0.1

The two just examined cases bound service loading and material response from below, and to the extent that ΔK_{th} = 2.6 ksi $\sqrt{in}$ is valid, from above as well. It is probable that much of the life in actual plants does not, however, experience the conservative extreme of continuous vibration at a 1 ksi range in materials which do not exhibit a threshold. As such, this limiting condition which shows virtually zero life is incompatible with service because the inherent assumptions are incompatible with service. But, it is likely that vibration will occur at magnitudes which approach or exceed 1 ksi during a part of the life of the plant. If this occurs late in the life, these results indicate that the large number of cycles which will exceed the threshold will cause very rapid crack growth.

The results presented suggest that if ΔK_{th} is low, vibration will give rise to crack growth from relatively small defects. Under these circumstances the growth is slow and, as long as no other large transients occur, there is adequate time to shut down, since cracks on the order of the assumed initial sizes can be detected in an ISI. Thus, the NDI industry must in this context reliably detect cracks of depth less than or equal to about 10 percent of the wall thickness.

    <u>Material Properties</u>. With regard to Table 1, the impact of
data such as that of Fig. 1 used in liu of the ASME water curve
is evident by comparing curves for the smaller initial flaw size
- curves 2, 4, 6, and 8 through 12 in Fig. 2. Of these curves
2, 4, 6, and 8 relate to situations where ΔK_{th} = 2.6 ksi $\sqrt{in}$ whereas
curves 9 through 12 are for ΔK_{th} = 0. Curves 9 and 10 are essentially identical because vibration effects without a threshold
dominate the crack growth behavior. However, when this effect
is not dominate, the composite curve shows a significantly reduced
life (i.e., less than 1 plant life as compared to more than 2 plant
lives) as compared to the ASME "water" curve. This is evident
in Fig. 2 as the difference between curves 2 and 6 as compared
to 4 and 8, and curve 11 as compared to 12. If indeed the composite
growth rate data reflect physical reality, the present ASME curve
is significantly nonconservative.

    The impact of material properties is observed in the present
context in the slope of the a versus N data. As evident in Fig.
2, the different material properties result in about a factor of
two difference in slope, a significant change. Clearly in the
case of the ASME curve, the results show that NDI must be capable
of reliably detecting cracks of a length equal to 10 percent of
the wall thickness. However, if the composite curve reflects reality, crack sizes significantly smaller will have to be consistently
detected.

    <u>Flaw Character and Orientation</u>. This subsection examines
the sensitivity of component life as a function of initial flaw
size, flaw aspect ratio, and flaw orientation.

    Consider first the influence of initial defect size on piping
system life. With reference to Fig. 2, note that the a versus
N curves originate from two distinct values of the ordinate. The
lower curves relate to the smaller initial defect size, the upper
curves to the larger size. Based on a comparison of these two
groups of curves, it is clear the initial defect size has a strong
influence on predicted plant life.

    Indeed, initial defect size may be one of the most significant
factors in controlling predicted life, since the smaller the defect,
the more shallow the a-N curve and the longer the life. This trend
is strongly nonlinear so that modest gains in detectability, i.e.,
reductions in the probability of nondetection at smaller defect
sizes, would mean substantial gains in life. Clearly, as long
as the probability of nondetection of defect sizes comparable to
those assumed does not approach near zero, the present results
suggest that the safety of some plant may be jeopardized. It is,
therefore, essential the NDI techniques advance to the extent that

one can say that the probability of nondetection of flaws of a depth equal to 10 percent of the wall thickness is near zero.

Consider next the sensitivity of plant life to aspect ratio, as presented for example in Fig. 3. Note that the same initial flaw depth applies to each set of curves. As shown in Fig. 3, piping life increases as the a/2c ratio increases and this trend is a strong function of the a/2c ratio, i.e., a 0.5 a/2c ratio equates to 27 plant lives and a 0.2 a/2c equates to less than 2 plant lives. Such a trend is, of course, to be expected because of the dependence of cracked area on a/2c since the cracked area increases with decreasing a/2c and the fact that the stress intensity factor, K, also increases as a/2c decreases. If the bias in cracked area did not enter this comparison, piping life would be less sensitive to a/2c. But, analysis of other cracked geometries where this bias is removed suggest that a/2c is still a significant parameter. As such, NDI techniques which image the defect shape are essential if the PSI and ISI's are to yield useful information.

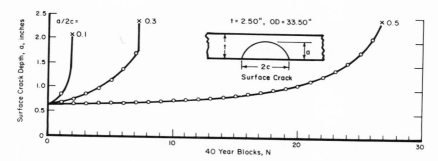

Fig. 3. Typical Results of the Sensitivity Study on a/2c for Circumferential Cracks

Finally consider trends with regard to orientation. Detailed analysis shows that the axial flaw consistently dominates failure of piping system components. For example, compare curve 1 of Fig. 2 with the curve for a/2c of 0.1 of Fig. 3, results which been generated under identical conditions except for crack orientation. In contrast to this analytical trend, service experience seems to indicate circumferentially oriented flaws are more frequent than axial flaws. In contrast to the fatigue growth rate process considered herein, this service experience is dominated by the incidence of stress corrosion cracks in boiling water reactors. Therefore, perhaps the difference between the predicted and observed dependence of crack development on orientation is due to the differ- ence in the mechanisms responsible for their growth.

Product Forms - Dissimilar Metal Welds and Nozzles. It is significant to contrast the difference between plant life for run pipes and elbows with that of dissimilar metal welds (DMW) and nozzles. In this limited study, one nozzle, and two DMW's, have been examined.

Consider first the DMW's, in the context of circumferential cracks over the range of initial crack sizes and aspect ratios detailed in Table 2. Again note that the pattern established here for the combinations of a_i and a/2c matches that for run pipe, as evident in comparing the curves shown in Fig. 4 with those in Fig. 2 and 3. However, the absolute life is very much less for the DMW as compared to run pipe in that the stresses local to the crack tip are higher than for run pipe[4]. Clearly then the DMW may be much more critical than run pipe or elbows. This is signifi-cant in the current context. It means that special care should be taken in the PSI and ISI's in this area. It also means that the threshold sensitivity of NDI techniques must be reduced in that growth from defects in such welds is very rapid. The inspection interval, therefore, must be short.

Table 2. Matrix for Sensitivity Study on a/2c and a_i for Circumferential Cracks in Dissimilar Metal Welds at Locations in AN-1 and SL-1

Curve[a] Number	a/2c	a_i Inches	a_i Inches
1	0.5	0.563	0.625
2	0.5	0.225	0.250
3	0.1	0.563	0.625
4	0.1	0.225	0.250

[a]The results are keyed to the curve number and plotted in Fig. 4.

Consider now nozzles,as detailed for part-through cracks in the matrix in Table 3. Curves developed for combinations of vibration/no vibration with threshold, a_i, and a/2c are shown in Fig. 5. Observe that the trends here are also comparable to the associated trends in Fig. 2 and 3 except that the life to through crack development is significantly shorter. Again, this is a conse-quence of the fact that again the stress field in the vicinity of the crack tip is higher than for run pipe[4]. As with the DMW, the nozzles appear to need special attention in both the PSI and ISI's. In particular, the frequency of the ISI should be increased as compared to run pipe. Likewise, because the growth rate is so high, special techniques which reliably detect smaller defects as compared to that used on run pipe should be adapted to the inspec-tion of nozzles.

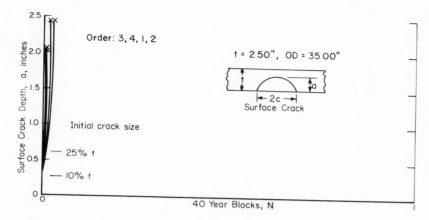

Fig. 4.   Typical Results of the Sensitivity Study on a/2c and a_i for Circumferential Cracks in Dissimilar Metal Welds

Table 3.   Matrix for the Sensitivity Study on Vibration With Threshold, Initial Flaw Size and a/2c for Nozzle Corner Cracks

Curve Number[a]	Vibration	a_i	a/2c
1	Yes	1.34	0.5
2	Yes	1.34	0.1
3	Yes	0.812	0.5
4	Yes	0.812	0.1
5	No	1.34	0.5
6	No	1.34	0.1
7	No	0.812	0.5
8	No	0.812	0.1

[a]These results are keyed to curve number and plotted in Fig. 5.

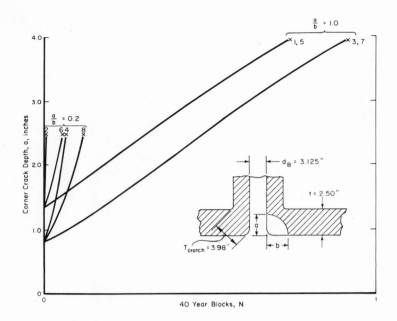

Fig. 5.   Results of Sensitivity Study on Vibration With Threshold,
          Initial Flaw Size and a/2c for Nozzle Corner Cracks

COMMENTARY

     Throughout the results presented it is evident that to ensure
plant safety by PSI and ISI's in the context of the damage tolerant
philosophy requires the reliable detectability of flaws less than
or equal to 10 percent of the wall thickness. In this regard,
it must be emphasized that statistics gathered on the successful
measurement of this size defect, while encouraging, tell us nothing
of those of the same size and larger that were missed. Since it
is the ones missed that grow, and since only one has to grow to
a critical size to cause a significant problem, it is essential
that cracks of this size be detectable with near zero probability
of nondetection. Unless this can be achieved in the field, by
field technicians, the fracture mechanics based damage tolerant
philosophy cannot be rationally used to assess piping system safety.
And, even if the probability of nondetection at such sizes is near
zero, the utility of the LEFM based approach in applications of
large-scale flow, as occurs in ductile piping materials, is questionable.

     Clearly, the propriety of the damage tolerant philosophy of
Section XI seems uncertain regarding both the inspectability of
piping and the viability of LEFM in piping applications. The open

question which arises in this regard is - can the more traditional
approach of Section III, which does not address initially present
defects in materials and workmanship, continue to be used until
the difficulties in implementing the damage tolerant approach are
overcome. Certainly the traditional approach has a large and successful experience base in pressure vessel and piping design.
But that base has been developed with materials and processes which
have been technologically superceeded. Furthermore, that technology
tells us little about the accumulation of fatigue damage in a fashion
which can be directly measured and correlated with service experience, until a failure occurs. Then it's too late. Neither can
that technology help decide if a cracked component can remain in
service, and for how long. As such it appears that a damage tolerant
approach is appropriate. Fracture mechanics must evolve to cope
with nonlinearities. But more significantly in the present context
NDI techniques should be further refined and special systems developed for the inspection of specific product forms.

SUMMARY AND CONCLUSION

     An assessment of the sensitivity of piping life to the character
of defects in piping systems has been presented following the development of the essentials of the damage tolerant approach. The
results are interpreted in terms of their implications regarding
nondestructive inspection. Specifically, this includes threshold
sensitivity, defect shape imaging, and defect statistics, including
the probability of nondetection.

     Significant conclusions that follow from the sensitivity analy-
ses include:
   • Threshold crack detection sensitivities on the order
     of 10 percent of the wall thickness are necessary to
     successfully implement damage tolerant concepts in the
     design of run pipe. Significantly lower thresholds are
     necessary for the successful application of these concepts
     other product forms.
   • The probability of nondetection must be near zero at
     or approaching the threshold detection level to ensure
     reliable designs based on damage tolerant concepts.
   • Special techniques and equipment should be developed
     and the inspection interval should be reduced for product
     forms other than run pipe (e.g., nozzles and DMW's) to
     insure low probabilities of nondetection and develop
     confidence in the application of damage tolerance concepts
     in piping system applications.

REFERENCES

1.   D. Broek, Elementary Engineering Fracture Mechanics, Noordhoff
     International Publishing, Leyden, The Netherlands, (1974).
2.   G. R. Irwin, Analysis of Stresses and Strains Near the End
     of a Crack Traversing a Plate, Transactions, Amer. Soc.
     of Mech. Eng., Journal of Applied Mechanics, (1957).
3.   H. A. Wood, and R. M. Engle, "Damage Tolerant Design Handbook",
     AFFDL-TR-79-3021, (1979).
4.   M. E. Mayfield, T. P. Forte, E. C. Rodabaugh, B. N. Leis, and
     R. J. Eiber, "Cold Leg Integrity Evaluation", NUREG/CR1319,
     (1980).
5.   "PVRC Recommendations on Toughness Requirements for Ferritic
     Materials", Welding Research Council Bulletin 175, (August,
     1972).
6.   W. H. Bamford, "Application of Corrosion Fatigue Crack Growth
     Rate Data to Integrity Analyses of Nuclear Reactor Vessels",
     ASME Paper Number 79-PVP-116, (1979).
7.   P. C. Paris, R. J. Bucci, E. T. Wessel, and T. R. Mager, "Exten-
     sive Study on Low Fatigue Crack Growth Rates in A533 and A508
     Steels", Stress Analysis and Growth of Cracks, ASTM STP 513,
     (1972).
8.   W. H. Bamford, and D. M. Moon, "Some Mechanistic Observations
     on the Crack Growth Characteristics of Pressure Vessel and
     Piping Steels in PWR Environment", Paper 222, Presented at
     NACE Corrosion/79, Atlanta, Georgia, (March, 1979).
9.   J. F. Kiefner, W. A. Maxey, R. J. Eiber, and A. R. Duffy,
     "Failure Stress Levels of Flaws in Pressurized Cylinder",
     Progress in Flaw Growth and Fracture Toughness Testing,
     ASTM STP 536, (1973).
10. M. F. Kanninen, et al, "Mechanical Fracture Predictions for
     Sensitized Stainless Steel Piping With Circumferential
     Cracks", EPRI ND-192, (September, 1976).

IN-FLIGHT ACOUSTIC EMISSION MONITORING

Gary G. Martin

Aeronautical Research Laboratories, Defence Science and
Technology Organisation, Department of Defence,
Melbourne, Australia

ABSTRACT

A miniaturised acoustic emission (AE) monitoring system has
been installed in a high performance jet trainer to monitor in-
flight growth of a known crack in a critical structural member.
It has been demonstrated that valid AE can be collected even in
this hostile environment. Spurious data from background noise have
been eliminated using a suitable frequency bandpass and a zone-
isolation technique. A linear correlation between valid AE and
increase in crack length, measured independently using a magnetic
rubber inspection technique, has been obtained. The AE data can
also be used to characterise the aircrafts' flight regime; most AE
occurs under rapid dynamic loading conditions typical of low flying.

INTRODUCTION

Acoustic emission has been successfully used as a nondestructive inspection technique in a wide variety of applications, including the detection and location of defects in structures such as
pressure vessels and bridges, and the detection and location of
corrosion and debond areas in bonded aluminium-honeycomb/aluminium
structures. Most applications have been to relatively large, on-
ground, stationary structures either in the field or in the laboratory where the space available for the complex monitoring equipment
has been relatively unrestricted. Hutton and Skorpic[1,2] of Battelle
Northwest Laboratories have recently produced a miniaturised AE
monitoring system, capable of discriminating between valid AE and
background noise, and suitable for use in a small high-performance
aircraft.

Most in-flight AE monitoring of aircraft has been conducted in North America. Mizell and Lundy[3] developed a system for the detection of unstable crack growth in C/KC-135 lower-centre wing-skin panels in-flight to insure "fail safe" operation prior to reskinning. Bailey[4] of Lockheed-Georgia had an active program to apply AE to the C5A transport. Rogers[5] conducted preparatory investigations into the application of AE to in-flight monitoring of the wing lugs of the F-105. Flight-noise analysis, laboratory specimen fatigue tests, transducer (sensor) placement studies and AE monitoring of ground fatigue tests on the wing main-frame forging were conducted, but the program terminated before in-flight monitoring commenced.[6] The Naval Air Development Center[7] plans to develop a monitoring system for routine in-flight application; initially the system will monitor the wing-carry-through box of the F-14 during full-scale fatigue testing. Bailey of Warner Robbins Air Force Base proposed monitoring the C-130 wing span.[7] McBride of Royal Military College, Canada is attempting to apply AE to the CF 100.[7] Although none of these projects have yet resulted in the widespread use of AE for in-flight structural integrity assurance, they have yielded valuable information.

This paper reviews the progress of tests being conducted in Australia on an RAAF Macchi MB 326 jet trainer aircraft (Fig. 1). This aircraft was selected for the project because the fleet is operating under a Safety-by-Inspection Life-Extension Program, in which the growth of cracks from known fastener holes in the centre-section assembly tension spar (Fig. 2) is being monitored by the magnetic rubber inspection (MRI) technique.

The MRI data is being compared with AE recorded during flights of the Macchi in normal squadron service. Progress results are presented and the future potential of AE for structural integrity assurance of airframes is discussed.

EXPERIMENTAL

Macchi

ARL and RAAF have undertaken a successful programme to extend the operational life of the RAAF fleet of Macchi aircraft beyond the initial design life, by allowing fatigue cracks to grow, up to a safe limit of size, in known critical locations of the tension spar of the centre-section assembly. The wings are attached to the aircraft via the large outer lugs of these spars which are made from 4340 steel heat-treated to the 1,100-1,240 MPa strength level. Cracking initiates in the walls of fastener holes 3 and 20 (Fig. 2), which are located near a change of section and are used to attach the air intake duct bracket to the spar. Fatigue cracks develop near the end of the design life and, for safety

Fig. 1. Macchi MB 326 jet trainer aircraft.

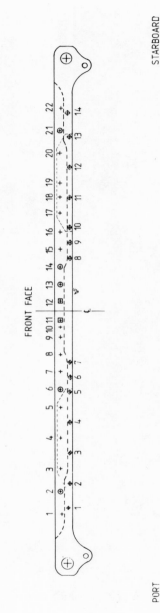

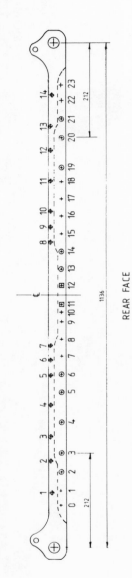

Fig. 2.   Centre-section assembly tension spar. Cracks initiate
          at holes 3 and 20 on the front face near the end of the
          design life. (Dimensions in mm).

reasons, the holes are reamed after another 1500 flying hours, when the estimated maximum crack depth (EMCD) falls in the range 1.38-1.57mm. The aircraft operate as military jet trainers and the particular aircraft being monitored is located at RAAF Base Pearce, Western Australia.

The crack growth is monitored every 100 hours using both an ultrasonic and MRI technique. A typical rubber plug, from which surface length and EMCD are determined, is shown in Fig. 3. These inspections require partial disassembly of the aircraft and each takes about two days to complete. Cracking in hole 20 of the selected aircraft was already well developed (EMCD=0.79mm), when in-flight AE monitoring commenced.

AE Equipment

The transducers and cabling were installed in June 1978 when the aircraft was undergoing a major service, and the preamplifiers and monitor unit were installed in August 1978. A schematic diagram of the equipment is shown in Fig. 4.

Transducer/Preamplifiers:

Tests conducted by ARL to determine the frequency and level of the background noise in a Macchi with the engine operating, indicated the suitability of a monitoring frequency of 400 kHz for the transducer/preamplifier system (Fig. 5). The transducers are tuned to reject all signals below 200 kHz, while the preamplifiers contain band-pass filters around a frequency of 400 kHz $\pm$ 25 kHz with a gain of 63 dB.

Two transducers, each 12.7mm diameter and 6.4mm high, were bonded to the web of the spar (Fig. 6) using a two-part acrylic adhesive (HYSOL-EA 9446). Flight trials by Lockheed Georgia[2] showed that this adhesive could withstand the required temperature and related dimensional changes in the metal spar.

Both preamplifiers are located on a shelf in the speed or airbrake bay (Fig. 7) close to the tension spar.

AE Monitor:

The monitor in the right-hand console of the front cockpit (Fig. 8) provides further signal gain (23 dB), processes and stores AE information, and provides power for the preamplifiers. The aircrafts' 115V AC supply powers the AE monitor unit.

System Operation:

Signal-processing involves sorting of signals on the basis of

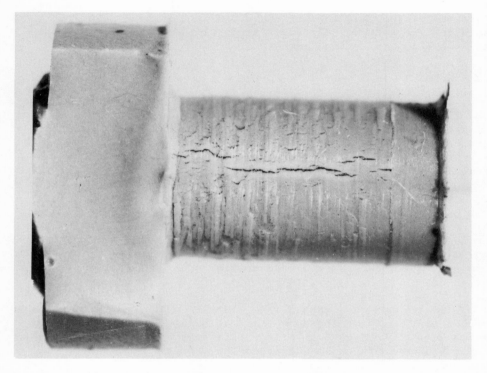

Fig. 3. Magnetic Rubber Plug taken from Hole 20 after 494.0 hours

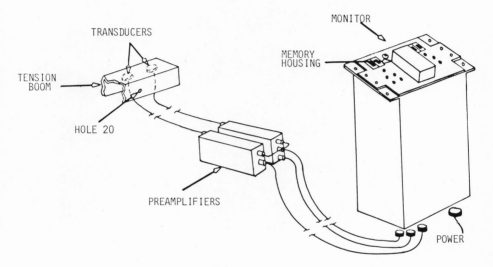

Fig. 4. Schematic of Onboard Aircraft AE System

10 dB

PER DIVISION

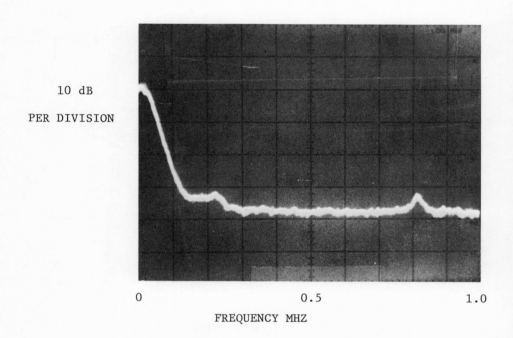

FREQUENCY MHZ

Fig. 5.   Noise Spectrum - Operating at Full Power

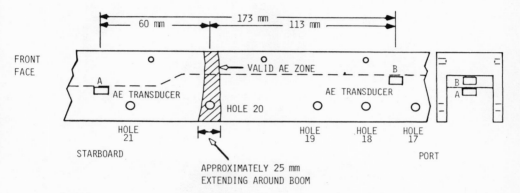

Fig. 6.   Location of AE Transducers and VALID AE Zone on Macchi
          Aircraft A7-021 Wing Centre Section Assembly Boom.

Fig. 7. Preamplifiers on Shelf in Speed-break Bay

Fig. 8. Monitor Unit in Cockpit Showing EPROM

their relative times of arrival at the two transducers (delta-time) so that only AE signals initiated within a zone of interest (isolation zone) bounded by two hyperbola are counted as VALID AE (Fig. 9). For each AE event whose amplitude exceeds a preset threshold level, a short logic pulse is produced. A signal from transducer A (which is closer to the isolation zone than transducer B) activates a timing window starting at t_1 and ending at t_2, all signals detected by B within the period $t = (t_2 - t_1)$ are accepted as VALID data. When an event is detected, a lock-out period of 4.1 ms is initiated, which prevents the same signal being processed more than once. Times t_1 and t_2 can be adjusted by means of thumbwheel switches on the monitor unit. The relationship between these thumbwheel settings and distances along the spar is shown in Fig. 9.

The AE data from each flight is stored in the electrically programmable read-only memory (EPROM) on the facia panel of the monitor unit (Fig. 8). Two types of data are stored: VALID AE, and TOTAL AE, i.e., all signals detected by the transducer array. There are 512 addresses for each type of data, each with a capacity of 61,440 counts. The memory is advanced to a new address every 1.1 minutes. A count is registered for each detected AE event whose signal exceeds the threshold level but information regarding signal amplitudes is lost during processing.

Data Collection

AE data is recorded on every flight and, to avoid overwriting of stored information, the EPROM is changed after each flight. However, as far as the pilot is concerned, the equipment operates automatically. The AE data is read out using the on-ground reader/printer (Fig. 10) to give a hard copy (Fig. 11). The memory is then erased under the UV source prior to re-use.

As well as the AE data from each flight, RAAF provide copies of fatigue meter readings (g exceedances) and flight profiles.

Improved methods of data-handling, to facilitate transfer of the information to a computer are presently being developed.

Equipment Calibration and Verification

Installation of the monitor unit and checking of the system (including four test flights) were accomplished in 4 days. After installation of the monitoring system in the aircraft, basic system functions were tested and isolation zone settings adjusted using an on-ground auxiliary power supply. Thereafter, the aircrafts' AC inverter provided power to check for possible noise problems related to engine operation, power supply, avionics or control system operation. No such problems were encountered on the ground and the system functioned satisfactorily. The first

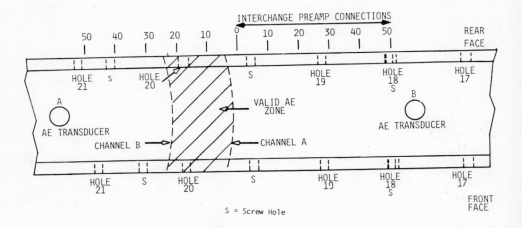

Fig. 9.   Thumbwheel Settings for VALID AE Zone

Fig. 10.   EPROM Reader/Printer and UV Eraser

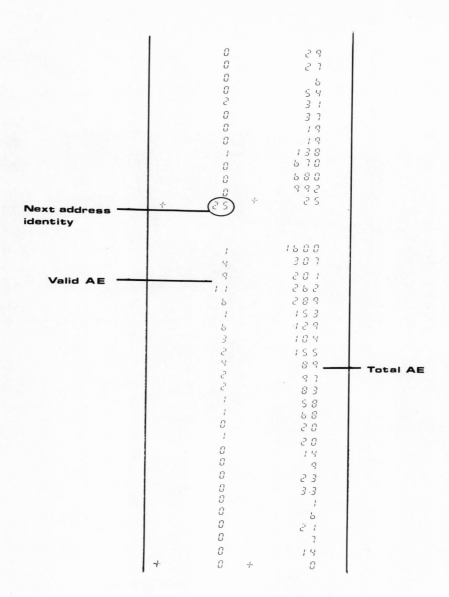

Fig. 11.   Hard Copy Output from Printer.
(VALID and TOTAL AE shown side
by side for convenience).

four test flights were used to check for the effects of in-flight
noise and transient signals, and to adjust the threshold setting.

A simulated AE source was subsequently used to verify the
relationship between thumbwheel settings and distances along the
spar (Fig. 9).

DATA ANALYSIS

Analysis of data from the first 61 flights has provided some
interesting results. Comparison of the VALID AE and TOTAL AE on a
flight-by-flight basis shows that the VALID AE does not follow the
same trend as the TOTAL AE, i.e., the VALID data is not simply obtained from uniform reduction of the TOTAL AE. Similarly, distribution plots (Fig. 12) indicate that large TOTAL AE counts can
coincide with VALID counts of zero and 1. A Spearman rank correlation analysis of all VALID AE and corresponding TOTAL AE (including
zeros in both sets of data) yielded an r value of 0.3. The same
analysis, excluding VALID AE zeros and corresponding TOTAL AE
values, yielded an r value of 0.6. These results not only support
the initial observation that the VALID data is not just a uniform
reduction of the TOTAL data but also indicate that most of the
VALID AE (counts $\geq$ 1) occurs when the TOTAL AE is highest.

Either set of AE data can be used to characterise the flight
regime of the aircraft, (Table I). At this stage, the characteri-
sation using TOTAL AE data is more striking (Fig. 13). AE activity
is lowest for formation-flying (which for our purposes here includes level flight above 610m (2000 ft), climbing to altitude,
and descent (other than in aerobatics). Activity increases for
circuits and landings, and increaes further for aerobatics although, in this regime, bursts of high activity interspersed with
quieter periods, are likely. The greatest activity occurs with
low-level flying, where the aircraft encounters buffeting or gust-
loading from ground thermal effects. In general, AE activity
increases as dynamic loading increases.

AE data from some flights have been lost but estimates of the
lost data have been made, based on knowledge of the relevant flight
profiles.

MRI was carried out just prior to installation of the AE
monitoring system and at approximately 100-hour intervals since
that time. The isolation zone includes both hole 20 on the front
face and hole 20 on the rear face. Hence, both holes are monitored
for crack growth, but no crack has been detected in the hole on the
rear face. All cracks, or crack-like indications, on both sides of
hole 20, e.g., Fig. 3, are measured after each inspection. The
increase in total crack length on both sides of the hole is given
in Table II together with the AE data (both recorded and estimated).

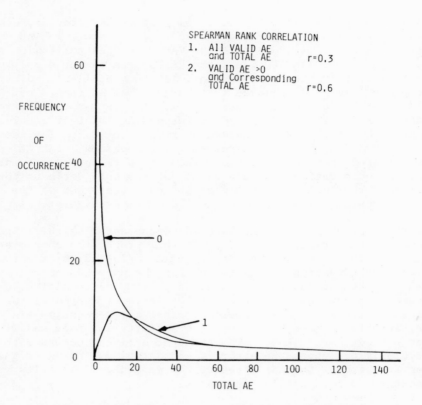

Fig. 12.   Occurrence of TOTAL AE Corresponding to
           VALID AE of 0 and 1.   (First 61 flights).

TABLE I.   DETAILED ANALYSIS OF AE FOR EACH FLIGHT REGIME

Flight Regime	Valid AE Average Counts/Hr	Total AE Range of Counts/ Address	Remarks
Formation	4	0 - 60	Mainly 0
Circuits and Landing	8	0 - 90	Some 0 Generally 30
Aerobatics	14	0 - 200	Low with High Bursts
Low Flying	163	5 - 250	Continuously High 40

TABLE II.   RESULTS OF 462.1 HOURS AE MONITORING

Hours	Flights Flown	Monitored	Total Valid AE Recorded	Estimated*	Crack Length Increase(mm)
0	0				1st MRI
31.9	24	0	0		0.16[ø]
105.2	62	62	1,843	1,843	0.50
206.3	79	18	2,462	4,549	0.71
303.8	81	61	4,168	6,817	0.96
399.0	80	58	5,230	8,347	1.36
494.0	81	71	7,246	10,651	1.75

* Estimated from flight profile data

ø Estimated value, start of AE monitoring

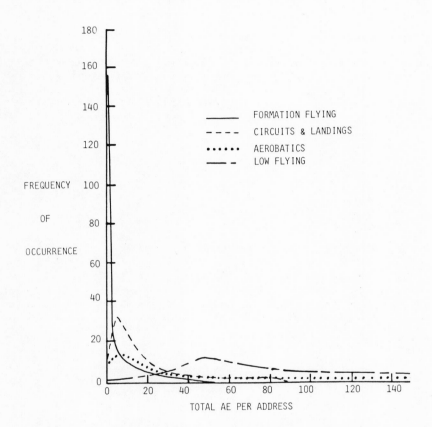

Fig. 13.   Distribution of TOTAL AE for 4 Flight
           Regimes. (First 61 Flights).

This information is summarised in Fig. 14, where a good relationship between increase in crack length and total VALID is shown.

DISCUSSION

Background noise from both electronic and mechanical sources
has been a major limitation in previous attempts at AE monitoring
under dynamic conditions. In the present project, the effects of
background noise were considerably reduced by the use of a selected
frequency window and zone-isolation techniques. However, the
effects of background noise cannot be eliminated entirely and could
generate some pulse pairs of time differences which give a VALID
count. Patterson[8] shows that errors can occur in the BNWL AE
monitoring system when noise sources are present on both sides of
the transducer array. A measurable estimate of the number of
erroneous VALID signals that can be expected during a typical
flight can be made as follows:

Consider two independent background noise sources α and β
outside the transducer array and generating noise at the same λ
average rate   according to a Poisson distribution. For a typical
flight lasting 4,400 seconds, in which 1300 TOTAL counts and 15
VALID counts were generated, it can be shown that the probability
of an erroneous VALID count being generated by pulses from α and
β is

$$P\ (E)\ =\ 1\ -\ e^{-\lambda\Delta\ t}$$

which approximates to

$$P\ (E)\ =\ \lambda\Delta\ t.$$

Now assume that all the TOTAL AE arises from α and β ; then

$$\lambda\ =\ \frac{1300}{4400}\ =\ 0.295$$

and, for $\Delta t\ =\ 14.1 \times 10^{-6}$ sec,

the probability of an erroneous VALID count is

$$P\ (E)\ =\ 4.17 \times 10^{-6}$$

As there are approximately 650 'pairs' of pulses from both α
and β, 2.71×10^{-3} erroneous VALID counts in this typical flight
would be expected.

The isolation zone was shifted to an uncracked, hole-free
region of the spar, by adjusting the thumbwheel settings, so as
to check the level of spurious valid counts. The total VALID

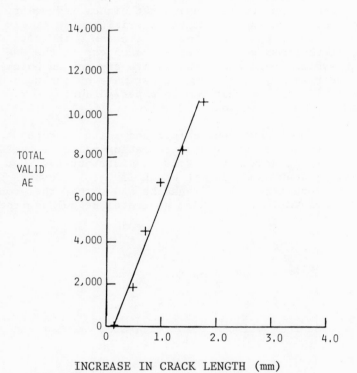

Fig. 14. Total Estimated VALID AE as a function of
          Increase in Crack Length.

count of 6, recorded over a period of 28 flights covering a representative distribution of flight regimes, was considerably below the total VALID count of at least 505 expected from the cracked region for a similar series of flights. Thus, the noise rejection techniques appear to be working satisfactorily.

Fretting, e.g. between the front face of the spar and the panels attached to it and between the bolt and the spar, could possibly give rise to AE. However, earlier AE monitoring of laboratory fatigue tests on a Macchi spar[9] showed that no significant AE was detected from regions of known fretting unaccompanied by crack growth, although AE was readily detected from cracked regions. Although the transducer frequencies were different (175 kHz in the laboratory study c.f. 400 kHz) fretting (if present) is not contributing to the VALID AE signals in the in—flight tests.

A straight—line relationship between the increase in crack length, measured by MRI, and the total VALID AE is found. Considering that the experiment is being conducted in an aircraft under normal squadron service conditions, the data scatter is acceptable. Very little research has been conducted into the relationship between AE and short cracks but Ryman and Blackwell[10] noted similar relationships between the crack length and total AE when plotted against load cycles in their study of crack initiation and growth in BSL65 aluminium alloys. There is scope for research into the relationships between surface crack length, crack depth and AE for small cracks under fatigue loading conditions to determine whether the relationship observed is not just fortuitous.

AE produced by fatigue processes is considered to be primarily a consequence of dislocation motion within the region of the plastic zone[11,12]. There will be a contribution from the crack growth itself, but where the crack-growth increment is very small, as in the case of very small cracks, this will only be a minor part of the AE recorded. As AE increases with increase in strain rate[13], the increase in AE with rapid expansion of the plastic zone under dynamic loading is expected. AE during low flying indicates that some damage, albeit small, does occur to structures in this situation.

FUTURE APPLICATIONS OF AE IN AIRCRAFT

Results to date in this program indicate that, with a suitably designed monitoring system, AE from defects growing under fatigue loading can be detected in an aircraft during flight. Thus, use of similar equipment to monitor inaccessible fatigue-critical areas appears possible in the near future. Once confidence in the interpretation of results has been achieved, the need for regular costly disassembly and down-time for inspection by conventional non-destructive techniques will be obviated.

AE can be also used to monitor large areas and it should be possible to develop a system for locating and monitoring growing cracks within the area under surveillance, not just for monitoring known defects. Hence, it is now possible to monitor full-scale fatigue tests of new aircraft in order to identify fatigue-critical regions. This should eventually lead to the use of AE-generated information for structural integrity assurance.

CONCLUSIONS

AE can be detected and recorded in the hostile environment of an aircraft in flight.

A straight-line correlation was found between a crack-growth parameter (increase in surface crack length) and the total number of VALID AE counts.

The production of suitably designed AE monitoring systems for structural integrity assurance appears to be feasible.

ACKNOWLEDGEMENTS

The AE equipment used in this experiment was designed and built by Battelle Northwest Laboratories under a contract funded by the Defense Advanced Research Projects Agency (U.S.A.) - DARPA. The extensive contributions of Phil Hutton and Jim Skorpik of BNWL are acknowledged.

The cooperation of the RAAF in this experiment is gratefully acknowledged.

The cooperation of Jack Ewen and Ron Evans of the Commonwealth Aircraft Corporation during the planning stages is also gratefully acknowledged.

REFERENCES

1. Hutton, P. H. and Skorpik, J. R., "AE Monitoring Simplified Using Digital Memory Storage and Source Isolation," Mater Eval, 35 (11), 1977, 55-60.

2. Hutton, P. H. and Skorpik, J. R., "Develop the Application of a Digital Memory Acoustic Emission System to Aircraft Flaw Monitoring," PNL-2873 UC-37, December 1978.

3. Mizell, M. E. and Lundy, W. T., Jr., In-flight Crack Detection System for the C-135 Lower Center Wing Skin, ISA ASI 76240, 1976, 254-258.

4. Bailey, C. D. and Pless, W. M., "Acoustic Emission Structure-Borne Background Noise Measurements on Aircraft during Flight," Mater Eval, <u>34</u> (9), 1976, 189–195, 201.

5. Rogers, J., "PRAM PROJECT – The F-105 Acoustic Monitor," Final Report USAF PRAM Program, July 1979.

6. Rogers, J., 2nd Annual AE In-Flight Monitoring Workshop AETC, Sacramento, CA, March 1979.

7. Carlyle, J., Ibid.

8. Patterson, A. K., Private Communication.

9. Martin, G. G., "Acoustic Emission Monitoring During Fatigue Testing of a Macchi Boom," Presented at 2nd Annual Acoustic Emission In-Flight Monitoring Workshop, AETC, Sacramento, CA, March 1979.

10. Ryman, R. J. and Blackwell, M. P., "An Acoustic Emission Investigation of the Initiation and Propagation of Fretting Fatigue Cracks in BS265 Aluminium Alloy under Constant and Random Amplitude Loading," Rep. No. FAT/130, ISSUE 2 FOR M.O.D. PE AIRCRAFT EST, FARNBOROUGH, September 1977.

11. Green, G. and Walker, E. F., "Acoustic Emission During Fatigue Crack Growth in AISI 4340 Steel," NDT 78, Aston University, U. K. 1978.

12. Green, R. E., Jr., and Pond, R. B., Sr., "Ultrasonic and Acoustic Emission Detection of Fatigue Damage," AFOSR TR-78-1284, July 1978.

13. Martin, G. G., "Effect of Strain Rate on Acoustic Emission from Aluminium Alloys," 9th World Conference on Non-destructive Testing, Melbourne, Australia, November 1979, 4J-4.

ON INTERRELATION OF FRACTURE MECHANISMS

OF METALS AND ACOUSTIC EMISSION

G. S. Pisarenko, S. I. Likhatsky, Yu. V. Dobrovolsky

Institute for Problems of Strength
Academy of Sciences of the Ukr.SSR
Timiryazev str., 2, Kiev 14,252014
USSR

ABSTRACT

A new method of nondestructive testing based on application of acoustic emission is discussed. The degree of correlation between AE characteristics variation, loading level and plastic deformation growth in materials during mechanical testing has been investigated. AE characteristics variation allows one to specify the intensity of local plastic deformations and to determine crack onset time. The AE method under consideration enables one to monitor the damage accumulation intensity in structural components when estimating their carrying capacity. All tests were conducted on both flat rod and single-edge notched specimens machined from different structural materials within a wide range of temperatures and deformation rates.

INTRODUCTION

Fracture of materials is known to be a multistage kinetic process of crack initiation and crack growth accompanied by mechanical, thermal, sonic, ultrasonic and other effects. Application of non-destructive methods to mechanical tests depends, to a great extent, on the development of new methods of monitoring the structural damage accumulation and the development of microplastic deformation, micro- and macrocracks and inherent or initiated fracture sources. The AE technique does have several advantages in this respect over other methods and its use as an efficient physical method for evaluating the state of solids under loads attracts growing interest of re-searchers.

Acoustic emission (AE) is a process of a spontaneous release of

367

elastic vibration energy during elastic and plastic deformation or initiation of micro- and macrocracks in materials. AE has long been in use for nondestructive tests [1,2,3,4]. However, it differs from other nondestructive techniques in that it is passive and thoroughly dependent on internal stress changes in the specimen.

AE caused by elastic vibrations is detected by one or several sensors which are mounted on the surface of the investigated material or structural component. The AE sensor follows the motion of the surface, converts it to electric signals which are then amplified, processed and recorded.

Prospects of the use of AE methods for deformation and flaw detection in materials are provided by its ability to directly monitor the damage accumulation process in the material with low elastic-plastic or high plastic deformations. The AE method is also advantageous for predicting the failure in metals under increasing or constant loads from the time of transition of damage accumulation to the critical state of a rapid fracture process.

The main objective of the preset study was to accumulate data on AE characteristics relating to mechanical tests of materials and structural members.

Investigations were conducted with the aim of establishing relationships between deformation and fracture processes and AE characteristics relating to tests of different materials under different conditions (with regards to strain rate, temperature, etc).

MATERIALS AND EXPERIMENTAL PROCEDURE

Special equipment for detecting, amplifying and conditioning AE signals as well as new experimental techniques has been developed at the Institute for Problems of Strength, Academy of Sciences of the Ukr. SSR. Basic parameters of AE were selected. The identity of AE diagrams was considered to evaluate the degree to which the equipment was free of interference and to check the performance reliability of the system for detecting, amplifying, and recording conditioned AE signals.

The fact that identical acoustic emission diagrams were obtained for a set of specimens of a metal tested under similar conditions as well as for the specimens of the same set tested in different tensile-testing machines testified to a high performance of the equipment employed and of the validity of acoustic stress-strain and fracture diagrams [5,6].

Widely-used structural materials of various classes were chosen for the experiment including aluminum, titanium and heat-resistant alloys, different steels, model material, glass, pyroceramics, composite materials, coal ore and some other materials.

To determine the physical nature of acoustic emission, the de-
pendences of the acoustic emission parameter on the material structure and some other factors were studied. The investigations were
carried out on specimens of zone-melted high-purity iron with 0.05%
C and commercial purity iron with 0.5% C with grains of various size.

The specimens tested were manufactured from the same batch of
each material. Geometric dimensions of flat smooth tensile-specimens
(gauge length of 3 x 10 x 70 mm) and specimens with a single-edge
notch (3 x 60 x 100 mm) were in conformity with the USSR standards.
Single-edge specimens with notches (proposed by Kahn) had the radius
of curvature at the notch root of 0.003 mm, 0.3 mm and 1 mm. Specimens were notched mechanically with an endless flat strip. Each
specimen in a set comprising at least 5 specimens was tested under
the same conditions.

Flat smooth specimens made of the materials under investigation
were tested in static tension in the temperature range of 20°C to
-196°C, with strain rates being 2.37×10^{-2} sec^{-1} to 4.76×10^{-4}
sec^{-1}. Flat notched specimens were tested in static uniform tension
and in tearing. The moment of the crack growth onset was determined
by two methods: visually with a MBS-1/x100/microscope - point 1 and
by the known method of conditioning fracture diagrams in conformity
with ASTM Standard - point 2.

In tests on notched specimens, the "load-crack-tip opening displacement" diagrams were processed using the above mentioned ASTM
technique [14]. To determine the relation between the crack length
and the crack tip opening displacement the K - calibration method was
used [15,16].

All tests were carried out in standard testing machines (UMM-20,
ZD-10, TT-KM-25 Instron).

Investigations involved utilization of strain and load measuring
units, a system for detecting, amplifying, conditioning, and recording the acoustic emission signal. A control unit allowed simultaneous switching on of all channels for data recording.

The tests were preceded by the following preparatory operations.

A specimen was fixed in the testing-machine grips by means of
pins. A strain-gauge extensometer for measuring strains or crack-tip
opening displacement was mounted on the gauge length of the specimen.
The acoustic emission detector was positioned on the upper elongated
end of the specimen.

Figure 1 shows a block diagram used for monitoring acoustic emission during testing at the Institute for Problems of Strength of the Academy of Sciences of the Ukr.SSR.

When loading the specimen a signal from the acoustic emission sensor is fed into the preamplifier with the gain K_u of 1000, the frequency band-width f of 10 to 6000 kHz and the noise level reduced for the input of U_{in} = 3 mV.

From the preamplifier the signal is received by the band-pass filter with the transmissions band f_n of 40 to 2500 kHz. The filters cut off vibration noise from the testing machine and decrease considerably the effect of external interference sources as well as basic noise of the equipment.

Then the signal is amplified and recorded by the wide-band tape-recorder in the frequency range of 10 to 3000 kHz. Simultaneously, the amplified signal is fed into the pulse-height analyzer and into the second channel of the pulse-frequency counter. The multichannel pulse-height analyzer AI-128-2 measures statistical pulse-amplitude distributions and transfers the accumulated information to the internal oscillograph and the digital printer of B3-15M type. The accumulated information can also be transferred to the x-y plotter in analog form.

From the amplifier output, the signal is fed into the detector for discrimination of the envelope. The direct component of the envelope is filtered and recorded by the oscillograph. On passing through the active filter the high-frequency alternating component is recorded by the low-frequency tape-recorder. Simultaneously, the signal can be viewed on the oscilloscope. From the active filter output, the signal is transferred through the buffer stage to the input of the Schmitt's filp-flop type signal conditioner with an adjustable operation threshold.

From the output of the signal conditioner, square-wave pulses are fed into the first input of the pulse frequency counter. Thus, in the described system the pulse frequency counter measures, and the oscillograph records, two intensities: intensity (frequency) of the acoustic emission signal envelop and the intensity of the unrectified signal. Each 5 seconds the pulse counter and x-y plotter record the sum of acoustic emission pulse envelopes exceeding the threshold. This information is of special importance when loading rates are low and when materials with low level of acoustic emission are tested.

It should be noted that a sufficiently fast response of all the measuring and recording equipment of the system is required since acoustic emission signal parameters vary rapidly rather than gradually

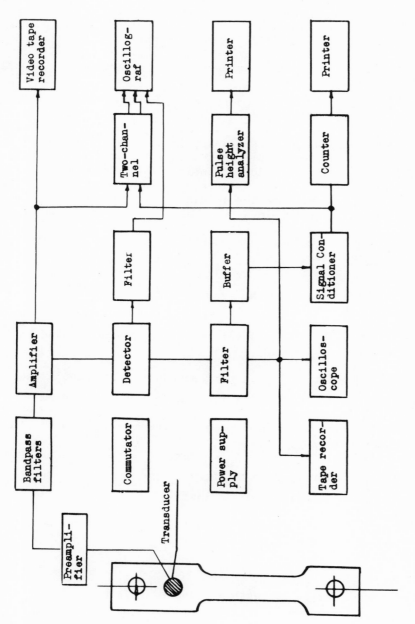

Fig. 1. Block diagram of acoustic emission amplification, measurement and recording system.

and are of random nature.

    In the course of the test, the variation of load (pressure), strain ($\Delta \ell$, mm), AE amplitude (U mV) and AE intensity ($\dot{N}$, pulse/sec) with time was recorded by an H-115 oscillograph.

    The processing unit made it possible to record the alternating component of the AE amplitude with the frequency of 0 to 15×10^3 Hz (U mV curve) and the signal proportional to the AE intensity in the range of 0 to 3000 pulse/sec ($\dot{N}$ pulse/sec) curve.

    Strain and stress produced in specimens or structural members under loading were determined from the oscillograms recorded with help of the strain gauges using acoustic signal diagrams. The variation of AE amplitudes (U mV), AE intensity ($\dot{N}$, pulse/sec) and total AE count (ΣN) with time from the loading start to the failure of specimens tested was determined.

    Figure 4 shows the joint stress-strain and acoustic signal diagrams, the following representative points were singled out on the time scale.

    Zero ($\tau = 0$) refers to the start of loading and deformation of the specimen.

    The first point (τ_1, sec) refers to the moment when AE amplitude reaches its first maximum (U'_{max}, mV).

    The second point (τ_2, sec) corresponds to the established standard for determining 0.2% offset yield strength ($\sigma_{0,2}$ kg/mm^2).

    The third point (τ_3, sec) refers to the moment when the AE intensity reaches its first maximum ($\dot{N}'_{max}$, pulse/sec).

    The fourth point (τ_4, sec) refers to the second maximum of the AE amplitude on the diagram (U''_{max}, mV).

    The fifth point (τ_5, sec) refers to the second maximum of the AE intensity on the diagram ($\dot{N}''_{max}$, pulse/sec).

    The sixth point (τ_6, sec) corresponds to the maximum load or to the moment when tensile strength is reached (σ_B, kg/mm^2).

    The seventh point (τ_d, sec) corresponds to the moment of the specimen failure and is the final point for the AE and stress-strain diagrams recorded.

Thus the portion of the stress–strain diagram between zero and τ_2 is referred to as elastic. The portion corresponding to strain hardening and the development of residual plastic deformation is located between τ_3 and τ_6. A necking of the specimen occurs in the portion of the diagram located between the points τ_6 and τ_d.

AE amplitude and intensity and respective values of stress and limiting relative strains were determined for these points. The numerical values were tabulated for each material tested.

RESULTS

The analysis of joint stress–strain and AE parameters diagrams made it possible to obtain quantitative information on the nature of the process of deformation and fracture of iron of different composition (Fe, α-Fe) with different grain-size and under various test conditions.

A pronounced AE amplitude and intensity maximum can be observed on all the diagrams as well as the largest increase in the total AE count covering the region of microflow, beginning with (0.6–0.9) σ_s. In this case the peak values of the AE amplitude and intensity in the iron of commercial purity is somewhat higher than in the material free of impurities.

It is shown that the decrease in the grain size, the increase in the strain rate and lowering the test temperature (Figs. 2, 3) cause a drop in plasticity and an increase in flow stress of the materials tested. At the same time, the values of AE amplitude and intensity grow considerably. This phenomenon is associated both with the flow stress increase and with the tendency of the material towards non-uniform deformation over the length of the specimen [7].

The analysis of joint stress–strain and AE parameters diagrams obtained from tests on specimens of AMg-6 alloy and Kh18H10T steel allows to define a number of characteristic features in the deformation process at 20°C and at −196°C.

Let us consider characteristic features of acoustic emission diagrams for each material.

AMg-6 alloy.

Figure 4 shows representative AE signals and stress–strain diagrams for specimens of AMg-6 alloy. AE diagrams exhibit two maximum amplitudes and two maximum intensities. The first maximum of the amplitude (U'_{max}) reflects a drop in the intensity. Then as the load is increased the first maximum of the intensity ($\dot{N}'_{max}$),

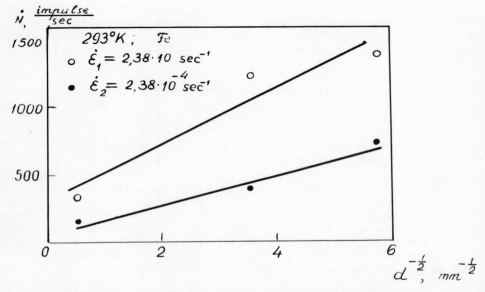

Fig. 2.   Acoustic emission intensity variation versus grain size.

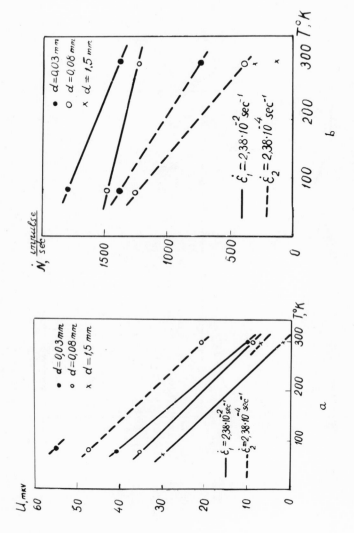

Fig. 3.   AE amplitude (a) and intensity (b) versus temperature
in commercial iron.

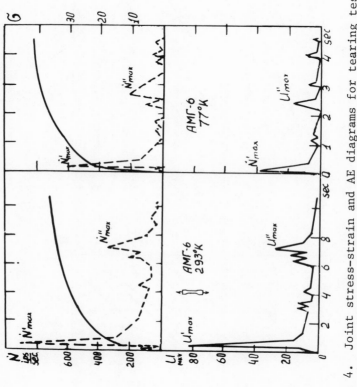

Fig. 4. Joint stress-strain and AE diagrams for tearing tests
of AMg-6 alloy flat rod specimens.

appears, being the largest of all the recorded ones.

At 20°C, the peaks U'_{max} and $\dot{N}'_{max}$ corresponding to the time point τ_1 and τ_3 precede the moment when the 0.2% offset yield strength $\sigma_{0,2}$ is reached (time τ_2).

Thus, the AE amplitude peak (U'_{max}) which corresponds to a drop in the intensity at the point τ testifies to the beginning of plastic deformation (early detection). Then, with the growth of load by 5-10%, the AE intensity peak ($\dot{N}'_{max}$) follows. The value of proof stress, determined from the stress-strain diagram exceeds stresses corresponding to the first amplitude and intensity maximum on the average by 1.8 kg/mm^2 and 0.9 kg/mm^2, respectively.

As the amplitude reaches its maximum, the intensity decreases abruptly and then increases with some drops in magnitude, which testifies to the non-uniform development of strain hardening of the polycrystalline material. All diagrams clearly reveal the second AE amplitude and intensity maximum, U''_{max} and $\dot{N}''_{max}$, after which both the amplitude and the intensity decrease up to the failure of the specimen.

The stresses corresponding to the points τ_4 and τ_5 are 93.4 to 97.9% and 91.4 to 96.9% of the tensile strength, respectively, for all three strain rates at room temperature. At -196°C, these stresses are equal to 91.4 to 96.6% and 88.3 to 96% of σ_B (Fig. 6).

The shape of the AE amplitude and intensity curves do not depend much on the temperature and strain rate while numerical values of these AE signal parameters increase as strain rate increases and decrease as temperature decreases.

Kh18H10T steel.

In specimens loaded at room temperature, AE amplitude increases sharply and reaches its first maximum value at stresses of 32.4, 45.1 and 55.7 of $\sigma_{0,2}$ for strain rates of 1, 10 and 20 cm/min, respectively. Then a sharp drop in amplitude is observed. A further increase in stressing up to ultimate tensile strength gives rise to another two peaks in amplitude equal to 0.3 to 0.7 U'_{max}. At stresses about 0.92 σ_B the amplitude begins to rise and reaches the magnitude of the U'_{max} order as fracture occurs.

On the whole the AE intensity plot duplicates the amplitude diagram though it has a more pronounced "saw-toothed" appearance.

Stresses corresponding to the first AE intensity peak $\dot{N}'_{max}$ are smaller than proof stress by 4, 6.8 and 8.3% for the three strain rates respectively. The second intensity maximum peak corresponds to fracture of the specimen.

At −196°C the appearance of stress-strain diagrams and AE parameters sharply changes. In the stress-strain diagram a fairly pronounced yield plateau is clearly seen. The first amplitude maximum peak occurs at stress by 20-30% lower than 0.2% offset yield strength. The next amplitude peak falls on the beginning of the yield plateau. After a sharp drop the amplitude steadily rises again and reaches its second maximum under loads of 91.6 to 97.5% of σ_B.

The low temperature AE intensity diagram has a fairly pronounced first $\dot{N}'_{max}$ peak with a half-way drop corresponding in time to the first amplitude peak. There is another peculiar amplitude peak at the beginning of the strain hardening region. After that peak, the AE intensity varies only slightly until the second $\dot{N}''_{max}$ peak is reached where stresses are 93.8-93.4% of σ_B.

Qualitative character of the amplitude and intensity curves does not change considerably, though AE numerical parameters increase as much as 1 to 1.5 times with strain rate. At low temperatures, AE parameters increase numerically 2-3 times for each of the three strain rates. This regularity was also observed for other materials investigated.

Variations in the first amplitude maximums U'_{max} and AE intensity $\dot{N}'_{max}$ versus strain rate at 20°C and −196°C is shown in Figs. 5 and 6. At higher strain rates the values of the amplitude and AE grow. Thus, AMg-6 alloy appears to be "noisiest" at a room temperature while Kh18H10T steel does that at lower temperatures.

When analyzing joint diagrams obtained in tests on flat single-edge notched specimens for all tested materials and temperatures one can observe the amplitude growth with loading the specimen during the elastic crack opening (Fig. 7).

In the range of transition from elastic notch opening to plastic flow at the notch root the first AE amplitude peak (U'_{max}) is observed. Further increase in loading causes AE amplitude decrease.

In all materials tested at 20°C in this investigation, AE intensity $(\dot{N})$ begins to increase at the start of specimen loading. When the temperature of the test becomes as low as −196°C, the intensity growth begins to shift along the time axis to higher time magnitudes.

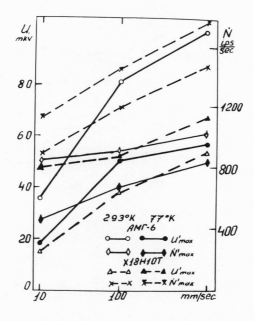

Fig. 5. AE amplitude and intensity versus strain rate.

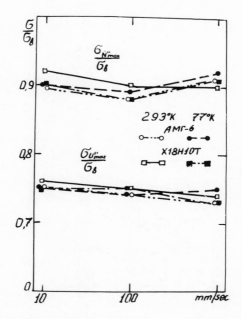

Fig. 6.   Relationship of stresses corresponding to U''_{max} ($\sigma_{u'' \, max}$)
and $\dot{N}''_{max}$ ($\sigma_{\dot{N}'' \, max}$) to tensile strength versus strain rate.

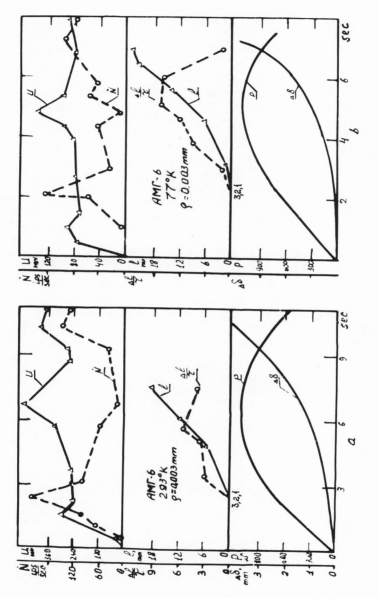

Fig. 7. Joint fracture and AE diagrams for AMg-6 alloy flat notched specimens.

This shift is due to the fact that as the resistance to straining increases with lower temperature, the development of plastic deformation is constrained and its onset occurs at higher loads.

Analyzing the AE intensity curves on fracture diagrams one can see that the AE intensity peak ($\dot{N}_{max}$) coincides with the crack onset point which was determined both visually and analytically (point 1 and 2). The moment of maximum intensity ($\dot{N}_{max}$) in the plots corresponds to point 3. After reaching its maximum value the AE intensity decreases in all the materials tested with only some slight rises and drops which can be attributed to non-uniform crack extension. AE intensity maximum corresponds to the highest load (Figs. 7, 8).

The maximum value of AE intensity growth corresponds to the crack onset time.

With account for a difference in loads when fixing crack onset time by the three techniques, visual, graphic (based on ASTM method) and acoustic, a possible discrepancy between stress intensity factors K has been established. The data in Table 1 indicate the load difference does not exceed 3–5% of the mean value while the discrepancy in stress intensity factor is 10 to 15%. The acoustic method gives the lowest K value and, thus, appears to be a method for the earliest determination of the crack initiation time. Comparison of joint diagrams showed that the points corresponding to the first maximum peaks and AE intensity may be used to characterize the process of crack initiation at the notch root and to predict the crack initiation time. Time from the beginning of the loading $\tau_{U'max}$ and $\tau_{\dot{N} max}$ corresponds to these points. Figure 8 demonstrates variations of these values for specimens with various notch sharpness. It should be noted that variation of the notch tip radius from 0.003 mm to 1 mm results in an increase of the absolute value U'_{max} for the set of

Table 1

Material	T°K	$\tau_{,s.}$	$\tau_{2,s.}$	$\tau_{3,s.}$	K_1	K_2	K_3
AMr-6	293	2.53	2.34	2.22	102.6	98.7	96.1
	77	2.50	2.41	2.33	85.8	85.8	79.6
X18H1Dt	293	5.5	4.9	4.2	361	345	319.3
	77	4.6	3.4	2.8	299.2	278	267.1
cm.65r	293	2.78	2.61	2.33	215.6	197.5	188
	77	--	--	3.72	--	--	

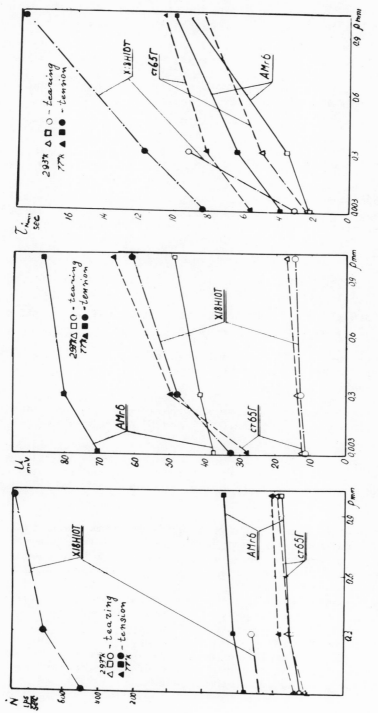

Fig. 8. AE parameters versus notch radius for different types of material and loading modes for flat notched specimen tests.

specimens tested regardless of the loading mode. In compact tension the U'_{max} value increases 1.3 times for AMg-6 alloy and 1.1 times for Kh8H10T steel and 65G-steel, while in uniform tension the increase is 1.5, 1.9 and 2.4 times, respectively.

The transition from compact to uniform tension for the set of tested specimens results in average increases of U'_{max} of 2 times for AMg-6 alloy, 3.7 times for Kh18H10T steels and 3.3 times for 65G steel.

The AE intensity maximum peak $\dot{N}_{max}$ of specimens tested in tear **rises** 1.5 times for AMg-6, 1.1 times for Kh18H10T steels, and 1.8 times for 65G steel as the notch sharpness is decreased, while in uniform tension these values are 1.1, 1.2 and 1.4 times, respectively. Tearing-to-uniform tension transition brings about average changes of 2.1 times for AMg-6 alloy, 3.4 times for Kh18H10T, and 1.1 times for 65G steel.

In tearing tests as the specimen notch is blunted the crack onset time ($\tau_{\dot{N}', max}$) increases 4.2 times for AMg-6, 3 times for Kh18H10T steels and 3.6 times for 65G steels whereas in uniform tension the increase is 2.5, 2.8 and 1.3 times, respectively.

Tearing-to-tension transition results in average time increase of 1.2 times for AMg-6, 2.3 times for Kh18H10T steel, and 1.2 times for 65G steel.

Thus, notch radius increase and tearing-to-tension transition results in the increase of the U'_{max} and $\dot{N}'_{max}$ values as well as in crack onset time delay (Fig. 8).

Figure 9 shows a joint diagram of pressure and the main AE parameters obtained in tests of a pipe structure with welded-in flanged joints.

Analysis of joint AE versus pressure diagrams shows that both AE amplitude and intensity grow in the first loading cycle of tested structures as the pressure begins to rise. Increasing pressure causes growing AE intensity. The more complicated the structure the higher is the amplitude and AE intensity for a given loading portion. AE amplitude changes with pressure in a monotonic manner while AE intensity curve has a time-dependent "burst" appearance.

When testing a notched structural component under increasing internal pressure the amplitude growth exhibits fairly pronounced jumps and drops along with smooth growth of AE intensity followed by a succeeding decline to its minimum values.

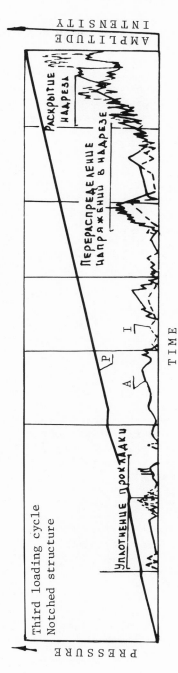

Fig. 9. Joint loading and AE parameters diagram for tests of a vessel with internal pressurization.

Further pressurization of the tested structure leads to a smooth
growth of the AE amplitude oscillating slightly, whereas AE intensity
sharply increases at the moments the amplitude drops down to the
least values.

With further pressurization the amplitude growth stabilizes
along with the sharp increase in intensity so that time–dependence
of intensity has a "burst"-like appearance.

DISCUSSION

Quantitative data on deformation and failure behavior of material
at low temperatures were obtained in testing the model material
(Fe and α – Fe). Clear relationships between variations of mechanical
properties (yield stress, tensile strength) and AE parameters were
observed. The flow stress was shown to increase with the decrease in
grain size. Since AE is caused, in fact, by dislocation movement, the
flow stress increase with the decrease in grain size at identical
values of plastic deformation will necessarily result in higher dislocation velocity, and, duly higher AE parameters (Fig. 2, 3).

An explanation for the increase in AE parameter with strain rate
can be given from the same standpoint [9].

AE intensity and maximum amplitude, as well as the highest increment in magnitude of all the signals covers the region of strain
curves between the yield peak and yield plateau. In iron, deformation
is known to occur within this very region by way of Chernov–Luders
slip-band propagation, and the bulk of deformation concentrates on the
front line of the band (dozens of microns thick) where the true strain
rate is several orders higher than the nominal strain detected on the
specimen [10,11]. This accounts for the above mentioned maximum magnitudes of AE parameters. The temperature decrease and higher impu-
rity content in the material was shown to cause both stress flow and
AE intensity growth in iron (Figs. 2,3).

When smooth samples of structural materials are subjected to
tension the volume of plastically deformed metal increases with
stress; i.e., plastic deformation starting in a site on the specimen
gauge length involves ever growing volume of the metal resulting in
AE amplitude increase. When the metal yield stress is reached (the
beginning of plastic zone formation), AE amplitude rises to its first
maximum value (U'_{max}).

When the specimen is loaded beyond the yield stress in the region
of intensive strain hardening, the AE amplitude decreases, AE intensity grows and reaches its maximum (N'_{max}) at the maximum damage
accumulation time in the points of the highest deformation.

If the specimen is notched, plastic deformation constraint occurs. The higher the stress concentration the more plastic deformation is localized. Therefore, the volume of the plastically deformed metal decreases and leads to smaller magnitudes of U'_{max} and $\dot{N}'_{max}$ (Fig. 8).

When the notch root radius increases, plastically deformed volume increases as well resulting in higher magnitudes of U'_{max} and $\dot{N}'_{max}$.

The loading mode, as seen in Fig. 8, also influences the plastic deformation progress at the notch root and fracture initiation.

The experiments revealed that the load corresponding to the moment of the crack growth onset under conditions of compact tension was much lower than that under uniform tension. If we assume that stress intensity at the crack tip is independent of the loading mode, then from the condition of equality of the stress intensity factors at the crack tip under both tension and tearing:

$$K_{tension} = K_{tearing}$$

where

$$K_{tension} = \sigma_{tension} \sqrt{\ell} \left\{ \frac{w}{\ell} \left[7.59 \left(\frac{\ell}{w}\right) - 32 \left(\frac{\ell}{w}\right)^2 + 117 \left(\frac{\ell}{w}\right)^3 \right] \right\}^{1/2}$$

$$K_{tearing} = \sigma_{tension} \sqrt{\ell} \left\{ \frac{w}{\ell} \left[505 \left(\frac{\ell}{w}\right)^3 - 236 \left(\frac{\ell}{w}\right)^2 + 49.4\left(\frac{\ell}{w}\right) \right] \right\}^{1/2}$$

(when $\ell = 20$ and $w = 60$), it is possible to obtain the ratio of nominal stresses $\sigma_{tension}$ and $\sigma_{tearing}$ acting in the cross-section not weakened by the crack, at the onset of its growth. In the case under consideration $\sigma_{tearing}/\sigma_{tension}$ ratio equals to 0.58. Since the area of the section not weakened by the crack is the same in both cases, the load ratio will be $P_{tearing} = P_{tension} = 0.58$ [14].

The variation of the specimen loading rate was insignificant with the crosshead speed being constant. Therefore, it can be assumed that loading was proportional to the time of deformation of the specimens; i.e., the time for loading the specimen up to the moment when fracture from tearing begins, will be approximately 0.58 of the time needed for pulling the specimen to fracture which is well confirmed by the experimental data presented in Fig. 8.

Considerable increase in U'_{max} and $\dot{N}'_{max}$ values with transition from one mode of loading (tearing) to another (tension) is, probably, associated with the increase in the deformed volume of the specimen at the notch root. The ratio mentioned above ($\sigma_{tearing}/\sigma_{tension} = 0.58$) is for all of the materials investigated (AMg-6 alloy, 12Kh18N10T and 65G steels) irrespective of the variation of the radius at the root of the notch on the specimen (from 0.03 to 1 mm).

Thus, in the course of tests on notched specimens, the absolute values of U'_{max} and $\dot{N}'_{max}$ vary with notch radius growth depending on the loading mode while the time of their attainment from the beginning of loading increases.

The data obtained allows one to recommend the use of AE parameters for the evaluation of the effect of various factors upon the moment of the crack growth onset and the process of its propagation when testing flat notched specimens fabricated of sheet metals.

The acoustic method can be used to determine the crack propagation rate in a sheet metal.

The results of the present investigations indicated that in loading specimens with a preinitiated flaw (a sharp notch, a precrack) the crack growth onset as well as its "stepwise" extension could be easily detected via the AE intensity behavior ($\dot{N}'_{max}$) (Fig. 8).

The data presented here on the variation of K with the value of the crack propagation rate agree with Hartbauer's results obtained in tests on L6AC steel specimens with a single-edge crack [17]. The crack propagation rate from the moment of its growth onset increases up to a certain maximum, then decreases by 40–50% and again increases until the moment of transition to the unstable mode of fracture.

CONCLUSION

1. A technique, along with an instrumentation system, has been developed which allows one to record an acoustic emission signal during elasto-plastic deformation in the material.

2. It is shown that it is advantageous to use the average amplitude value (U, mV) and the acoustic emission signal intensity as main parameters of the acoustic emission signal ($\dot{N}'$, pulse/sec).

3. It has been found for all the materials tested that the first absolute maximum of the amplitude (U'_{max}) preceds in time the moment when 0.2 offset yield strength ($\sigma_{0,2}$) is attained which, when testing specimens allows one to specify the moment of the gross plastic flow onset for metals that have no yield plateau. The existance of interrelation between the variation of acoustic emission signal parameters, the loading level and the development of elastic and plastic deformation in specimens has been determined for various structural alloys.

4. A conclusion has been drawn that on the acoustic signal diagrams certain points can be singled out which correspond to the increase in amplitude (U''_{max}) and intensity ($\dot{N}''_{max}$). The stresses at these points are equal to (71 to 78%) and (86 to 98%) of tensile strength.

5. It has been found for all the metals tested that variation in the acoustic emission signal parameters allows one to characterize the intensity of the development of local elastic deformation at the notch root and to define the moment of the crack growth onset. Thus, it becomes possible to obtain new data on the nature and regularities of ductile and quasibrittle crack propagation.

6. When testing complex structural components AE signals of a certain nature were detected which preceded the beginning of shrinkage and fracture of sealing elements and accordingly the beginning of leakage. In this case the leakage is detected much earlier from the AE parameters than by other methods.

7. It is shown that the use of the acoustic emission method makes it possible to diagnose the state of shell-structures, to define the nature of the damage of a structure, and to predict the initial stage of its failure or the deterioration of its performance.

REFERENCES

1. Ioffe, A. F., Physics of Crystals, Moscow-Leningrad, Gosizdat, (1929), p. 192 (in Russian).
2. Förster, F., Sheil, E., Akustische Untersuchung der Bildung von Martensitnadeln. - Z. Metallkunde, (1936), No. 9, s. 245-247.
3. Mason, W. P., McSkimin, H. J., Shockley, W., Ultrasonic Observation of Twinning in Tin, Phys. Rev., (1948), 73 (10), pp. 1213-1214.
4. Keiser J. Erkenntnisse und Folgerungen aus der Messung von Geräuschen bei Zugbeanspruchung von metallischen Werkstoffen, Archiv für das Eisenhüttenwesen, (1953), H 1/2, s. 43-45.
5. Likhatsky, S. I., Novikov, N. V., Acoustic emission indication technique for metal fracture processes, "Stroitelnaya Mekhanika Korablya," (1972) issue 173.
6. Likhatsky, S. I., Novikov, N. V., Voinitsky, A. G., Instrumentation and investigation of material physical properties, Naukova Dumka Publishers, Kiev, (1974) (in Russian).
7. Krasowsky, A. J., Novikov, N. V., Nadezhdin, G. N., Likhatsky, S. I., Correlation between acoustic emission, plastic flow and fracture in iron under static loading within a wide temperature and strain rate range, "Problemy

Prochnosti," (1976) No. 10, pp. 3-11.

8. Srawley, J. E., Brown, W. F., Mat. Standart, Vol. 7, No. 6, (1967).

9. Conrad, H., in: Ultrafine Grain in Metals, Moscow, Metallurgia Publishers, (1973) (in Russian).

10. Turner, A. P. L., Vreeland, T., Acta Met., Vol. 18, No. 11, (1970).

11. Klyavin, O. V., Regularities of Plastic Deformation of Solids in Liquid Helium Environment, Author's synopsis of doctor's thesis, Leningrad, Ioffe PTT, Academy of Sciences of the USSR, (1975) (in Russian).

12. Likhatsky, S. I., Novikov, N. V., Voinitsky, A. G., On recording stress wave emission in mechanic tests of materials. in: Nondestructive Inspection of Stress-Strain State of Structural Materials and Members by Stress Wave Emission, Abstracts of conference papers, Khabarovsk (1972).

13. Novikov, N. V., Likhatsky, S. I., Maistrenko, A. L., Determination of crack onset time by acoustic emission technique for notched compact-tension specimen tests, "Problemy Prochnosti, No. 9 (1973) pp. 21-26.

14. Boyle, R. W., A Method for Determining crack growth in Notched Sheet Specimens, Mat. Res. and Standards, Vol. 2, No. 8, August (1962).

15. Niemets, Y., Stiffness and strength of steel components, Moscow, Mashinostroyenie, Publishers (1970) (Russian translation).

16. Fracture 1969, Proceedings of the Second Conf. of Fracture (Brighton, 1969), London, Chapman and Hall.

17. Hartbower, C., Grimmis, P., Fracture of structural metals as related to pressure vessel integrity and n-service monitoring, "AIAA Paper," No. 501 (1968) p. 12.

FATIGUE DELAMINATION IN CFRP COMPOSITES : STUDY WITH ACOUSTIC EMISSION AND ULTRASONIC TESTING

F.X. De Charentenay, K. Kamimura and A. Lemascon

Département de Génie Mécanique
Université de Technologie de Compiègne
B.P. 233 - 60206 - Compiègne - France

INTRODUCTION

Non destructive techniques are useful not only as a quality control but also for the study of fracture mechanism. This is still more effective in the case of composites. For instance, these techniques are needed in the study of the phase of damages growth which very often precedes fracture in such materials. Several examples may be found in the published papers (1) (2) (3).

In fatigue of composite materials, damage growth is the key process of failure. If the stress is greater that the point of first ply failure, the fatigue life is the time elapsed between the first damage at the first cycle and the fracture at N cycles. At lower stress, fatigue in matrix may induce delayed cracking which is the initiation of damage growth. Thus the conditions for a safe fatigue behavior is at least to maintain the stress level below the point of first ply failure.

Knowing the importance of the matrix fatigue failure, we have studied the problem of fatigue delamination which is a shear fracture often encountered in a laminated composites. In order to avoid side effect, in plane shear failure or edges effect, which arises in a ± 45° or off axis shear tests, the well known short beam test has been used. In this paper we present a study of fatigue delamination by dynamic bending on Carbon-Epoxy unidirectional composites, with special emphasis on the use of NDT techniques, acoustic emission and ultrasonic testing.

EXPERIMENTAL TECHNIQUES

Materials and Samples

   Two Carbon-Epoxy composite with Thornel T 300 fibers were
studied : one with Epoxy Ciba 914 and the other with Narmco 5208.
The samples (120 mm x 20 mm x 5 mm) were cutted in 600 mm x 300 mm
or 300 mm x 300 mm plates. The plates were molded from 48 plies.
The fibers are parallels to the large dimension of the sample. All
the samples were checked before fatigue test with ultrasonic
(C.scan and attenuation measurements)

Mechanical Test

   The simple short beam three point bending test is not very accurate because stress concentration under the central loading nose.
In order to avoid this problem and to get a symetric dynamic stress
cycle we design a modified four point bending which is actually a
double cantilever test (fig. 1). The two shear zones are 12.5 mm
long which give a ratio of span width over thickness of 5. With
this ratio, fracture must occur in delamination by shear.

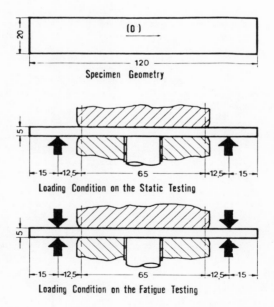

Fig. 1 : Double cantilever test module.

Dynamic tests are conducted on a closed loop servo hydraulic machine (MAYES 5 T) with stress control at 5 Hz. Maximum deflexion is continuously monitored with LVDT and a peak voltmeter.

Ultrasonic Testing

Samples are inspected before and after fatigue by C.scan in a water tank with focused piezotransducer (5 MHz). Ultrasonic instrument is a Krautkramer USIP 11. In the displayed C.scan the defects appear dark area. Attenuation coefficient (α) were also measured on eleven areas of the samples (including the two shear stressed zones) with a 10 MHz panametrics flat transducer.

Acoustic Emission

The acoustic emission system is made of a resonant transducer (250 KHz), a preamplificator (40 dB), a high pass filter, an amplificator (20 to 40 dB) and final results may be monitored as counts rate (discriminator with threshold level from 0.2 V to 1 V), as RMS or as maximum amplitude (B.K. 260). Special grips have been designed for lovering machine noise. The transducer is firmly fastened to the sample in the central zone where there are no sample deformation.

RESULTS AND DISCUSSIONS

Static Tests

Several samples were tested in a monotonic bending test. Results are given in table 1. However dispersion is fairly high. For the T 300 -N 5208 the influence of materials structure is shown by the relationship which appears in plotting the shear static strength τ_{st} in function of the attenuation coefficient of the shear stressed zone as was already demonstrated by Stone (4), Teagle (5) and Vary (6) (fig. 2). For the T 300.914 such relationship appears only if the attenuation coefficient of the shear stressed zone is normalized by the mean coefficient of the sample.

Delamination Description

Fig. 3 shows a typical experiment displaying A.E. (RMS) evolution versus cycles number. Although maximum amplitude is sometimes more sensitive, the RMS appears to be more reliable for the whole phase of delamination, initiation and growth. As shown by several experiments :

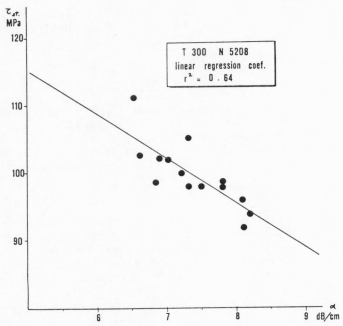

Fig. 2 : Attenuation coefficient influence on static strength.

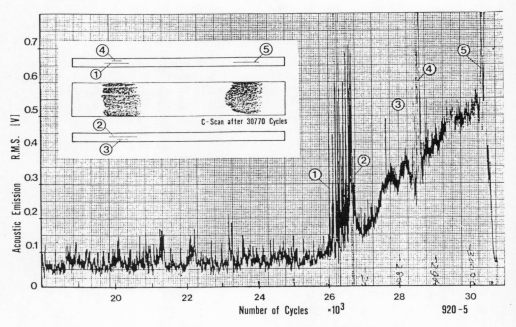

Fig. 3 : A.E. enregistrement during fatigue loading.

- Acoustic emission often occurs before any visual observation of crack on the edge of the sample.

- Acoustic emission is maximum when the crack emerges on the edge of the sample.

- Acoustic emission may detect multidelamination.

- The compliance increasing is observed a few cycles after A.E. has occured.

Figure 4 confirms the process of delamination in this test. A sample was observed by C.scan at a few given time of the fatigue life and the dark area evolution of the delamination shows that the mechanisms is well represented by the drawing of the figure 5. There are a stepwise lateral propagation of the delamination.

The carefull examination of the C.scan shows that the delamination does not start preferentially on a defect, say on a small dark area shown by C.scan. Optical microscopy on a starting delamination has confirmed that the fatigue crack ignores very often porosity.

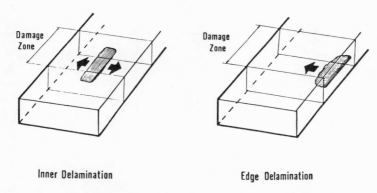

Inner Delamination   Edge Delamination

Fig. 4 : Shear fatigue damage initiation and propagation.

S.N. Curves

A.E. may be used as a precise point of first damage or of delamination initiation. Thus we use this criteria in order to define a S.N. curve which gives the number of cycles necessary to initiate delamination under a given maximum shear stress. Fig. 6 and 7 show these SN curves for T 300 - N 5208 and T 300-914.

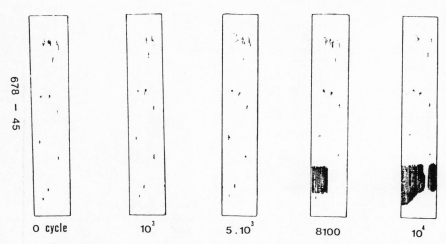

Fig. 5 : Delamination evolution during fatigue test observed by
         C.Scan
         T 300 – N 5208 materials
         Ultrasonic test : 5 MHz focused – 28 dB.

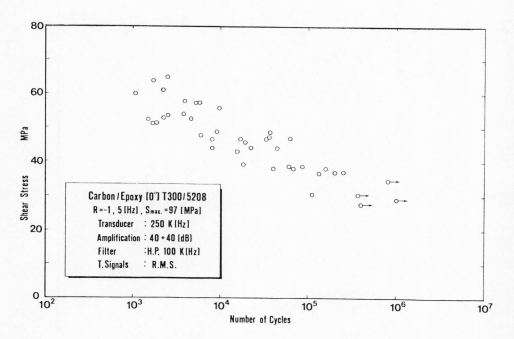

Fig. 6 : Shear fatigue S-N curve from acoustic emission.

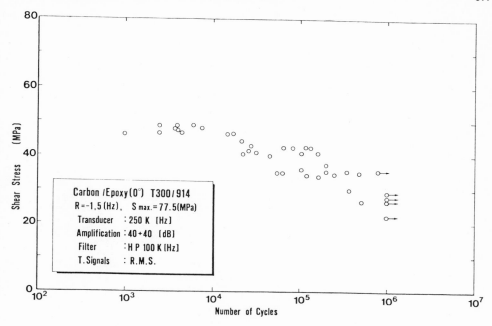

Fig. 7 : Shear fatigue S-N curve from acoustic emission.

The fatigue sensitivity is high for the T 300 N 5208. The T 300
914 although less resistant in a static test is as resistant as
the T 300·N 5208 for low stress level or long fatigue life (30
to 35 MPa). Dispersion is very high and as for many S.N. curves this
may be due to the lack of reproductibility of the samples.

Fatigue behavior and Attenuation Coefficient

In order to check the influence of materials parameters we
have plotted for a set of shear stress the influence of attenuation coefficient on the fatigue life. Results are presented
on fig. 8   for the T 300 N 5208. This influence is here clearly demonstrated. Higher is the attenuation coefficient, lower is
the fatigue life for a given maximum shear stress. And the influence of the attenuation coefficient is more important for the low
stress part of the curves, i.e. for the long fatigue life. Since
the attenuation coefficient may be lowered by a carefully designed procedure of molding and curing, this stage is of a first
importance for the fatigue life of the composite.

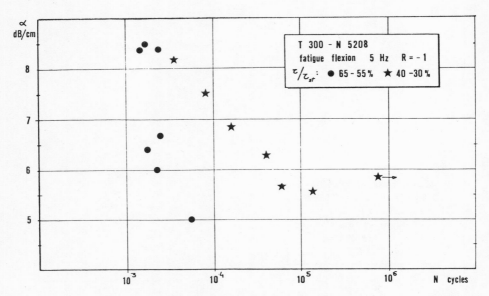

Fig. 8 : Influence of attenuation coefficient on fatigue life.

Normalized S.N. curves

Since the attenuation coefficient, not only influence static shear strength τ_{st} but also fatigue life, the S.N. curves may be well corrected by the materials parameters by using a normalized shear stress τ / τ_{st}. However τ_{st} cannot be, of course, determined before a fatigue test. Thus we use the plots of fig. 2 ($\tau_{st} - \alpha$) in order, from a measured α, to calculate an hypothetic τ_{st}. The normalized S.N. curve is presented in fig. 9 for the T 300 N 5208. The dispersion is lower than in figure 6 but is still rather high especially in the low stress zone. Thus the fatigue behavior is not simply related to the static mechanical strength and the materials parameters or other variables should have significative influence.

For the T 300 - 914 such a processing is difficult because of the use of normalized α coefficient and the differences between the mean α coefficient in the four plates used in the experiments

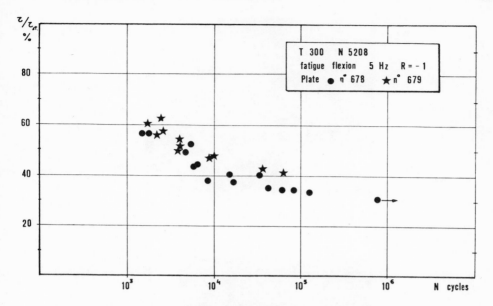

Fig. 9 : Normalized SN curve for T 300–N 5208 materials.

Fractography

        Analysis of delamination mechanism has been completed by op-
tical microscopy or SEM on polished section in the shear stressed
zone. Photograph 1 is an example of the step wise propagation of
crack from an interface to another in the T 300-N 5208.

        Fractography by SEM on delaminated surfaces (static or in
fatigue) are rather complicated. Photograph 2 for T 300·N 5208
shows what appears to us to be characteristic of shear fatigue
fracture. The T 300·N 5208 shows multiple microfracture transverse
to the fiber direction and with a spacing which is function of
distance from fiber to fiber (in delamination plane and out of
this plane). In T 300·914 (photograph 3) the nodular additives overcome almost all specific features of a given mode of failure.

        The photograph 4 is taken in a frontier zone which separates
a fatigue delamination zone and a monotonic delamination zone (obtained in a test of residual strength on a partially delaminated
sample). Clearly visible, the misalignement zone is one of the
probable initiation point. Such defect is often encountered in
unidirectional Carbon-Epoxy composites. Because of stress concentration in matrix at crossing fibers, this small zone may be the point
where, depending of the state of matrix (α, attenuation coefficient)
the first damage will occur and initiate delamination.

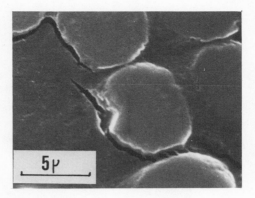

Photo 1 : Aspect of delaminage propagation.

Photo 2 : Shear fatigue fracture of T 300 - N 5208 materials.

Photo 3 : Nodular structure of BSL 914 resin.

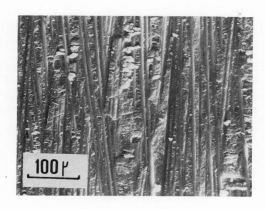

Photo 4 : Misalignement defect in frontier zone.

CONCLUSION

In this study of fatigue delamination in unidirectional
Carbon-Epoxy we have shown that the NDT contributions to the results were decisive.

- Acoustic emission is very useful for defining a point of
first damage during a fatigue experiment.

- C.Scan monitors delamination and confirms mechanism.

- Attenuation coefficient reveals the important influence of
materials parameters on the shear fatigue properties of Carbon-
Epoxy composites.

1. F. H. Chang et al, Application of a special X ray non destructive testing technique for monitoring damage zone growth in
   composite laminates, ASTM STP 580 (1974) p. 176
2. M. J. Salkind, Early detection of fatigue damage in composite
   materials, J. Aircraft, Vol. 13, n° 10, p. 764 (1976)
3. D. T. Hayford and E. G. Hennecke, A model for correlating damage and ultrasonic attenuation in composites, ASTM STP 674
   (1979) p. 184
4. D. E. W. Stone and B. Clarke, Non destructive determination of
   the void content in cfrp by measurement of ultrasonic attenuation, RAE TR 74162 (1974)
5. P. R. Teagle, Airworthiness certification of composite components for civil aircraft. The role of non destructive evaluation, 11th Nat. SAMPE Techn. Conf. (1979) p. 192
6. A. Vary and K. J. Bowles, Ultrasonic evaluation of the strength
   of unidirectional graphite polyimide composites, NASA TMX
   73646 (1977)